Foundations for Sustainability

Foundations for Sustainability

A Coherent Framework of Life–Environment Relations

Daniel A. Fiscus

Western Maryland Food Council, Cumberland, MD, United States

Brian D. Fath

Department of Biological Sciences, Towson University, Towson, MD, United States

and

Advanced Systems Analysis Program, International Institute for Applied Systems Analysis, Laxenburg, Austria

ACADEMIC PRESS

An imprint of Elsevier

Academic Press is an imprint of Elsevier
125 London Wall, London EC2Y 5AS, United Kingdom
525 B Street, Suite 1650, San Diego, CA 92101, United States
50 Hampshire Street, 5th Floor, Cambridge, MA 02139, United States
The Boulevard, Langford Lane, Kidlington, Oxford OX5 1GB, United Kingdom

British Library Cataloguing-in-Publication Data
A catalogue record for this book is available from the British Library

Library of Congress Cataloging-in-Publication Data
A catalog record for this book is available from the Library of Congress

ISBN: 978-0-12-811460-5

For Information on all Academic Press publications
visit our website at https://www.elsevier.com/books-and-journals

Publisher: Candice Janco
Acquisition Editor: Candice Janco
Editorial Project Manager: Emily Thomson
Production Project Manager: Joy Christel Neumarin Honest Thangiah
Cover Designer: Victoria Pearson

Typeset by MPS Limited, Chennai, India

Dedication

To our mentors: Bernard C. Patten and Robert E. Ulanowicz.

DAF: To my wife, Tracy Behan, who has taught me so much about connective tissue, family, love, hard work, and the benefits of a well-organized environment. You are truly the best, and you are the better half of our "coupled complementary process."

DAF: To my Dad, W. Guy Fiscus, who has taught me so much about sitting still outdoors to learn from and delight in the wonders of nature, about the arts and sciences of diagnosis, and the value of good questions.

BDF: To my family, Natasha, Maria, Peter, and, to all who have inspired me.

To our many students.

Contents

List of Figures

List of Tables

Preface

In this book, we make an ambitious claim—herein you will find scientific principles, methods, models, and ways of thinking that are uniquely capable for solving the chronic and systemic "humans in the environment" problem. While the impact for which we aspire is large, the creative credit and sources of insight we never claim as solely our own.

To take the broad and holistic view, to attempt to see, comprehend, and act wisely for both the life—environment and human—environment relations as undivided wholes, we have had to stand on the proverbial "shoulders of giants." Robert "Bob" Ulanowicz and Bernard "Bernie" Patten are the holistic science revolutionaries and trail blazers whose towering legacies of work have afforded us this wide-angle view. To have been taught and mentored by such visionaries, and to later have the humbling privilege to work side-by-side as colleagues, is necessarily to catch some of the infectious urge, excitement, dream, and action plan for revolutionary science in service to people, ecosystems, and living beings. If any of our ideas or proposals seem a stretch to comprehend, and if in searching to answer your questions you dig deeply into to the great bodies of work by Bob and Bernie, then this book will have served a good purpose. We have sought to present a comprehensive picture, but you will almost surely find their work necessary to help understand the full foundations of the science reform we describe.

In Chapter 7, Bridge Between Science and Applications a La Rosen, we retell a story and anecdote of a discussion between Bob and Bernie with respect to their differing views of the workings of Nature. In 1999, they debated the relative mechanistic and deterministic aspects of airplanes, an area and system they each know well as pilots. Their debate played out in the pages of the journal, *Estuaries*. The story is both fun and insightful, and we won't spoil it here. But extending the airplane analogy can serve to depict a strategic aspect of this book. If you have flown with a commercial airline, then you have heard the pre-flight safety *spiel*. The instructions always include the protocol on what to do "in the event of loss of cabin pressure," when the oxygen masks will drop down. The flight attendants insist to put on your own oxygen mask before helping others with theirs. This common-sense principle to ensure that your own "life support" is established prior to attempting to be of service to others links to a deeper principle that is wise in general. We employ an analogous principle in this book—we explain why science must learn to sustain itself before it can successfully promote sustainability in other sectors.

Bob and Bernie have contributed science that goes beyond Darwinian theory, and as we show, our present global ecological crisis calls for radical reform in the foundations of life science. When it first came out, Darwin's and Wallace's work, among many profound impacts, transformed people's sense of the human self in relation to the natural world. Evolution by natural selection presented a coherent framework by which to see ourselves as having coevolved with other species and as being descendants of prior species. Ecology, ecosystem and systems science, ecological networks, and sustainability science afford a similarly transformative new sense of self. As we strive to portray in this book, we humans are fully integral with the web of life "from origin to destiny." The original tapestry of life—environment relations we did not weave, but in the anthropocene earth, we are now very much responsible for major and creative aspects of the design and implementation of the planetary tapestry that is the web of life. We are at a fork in the road, and the choices and eventual destiny of life are our responsibility.

The new self-realization required from full understanding and appreciation of holistic ecology is equally as compelling as the Darwinian implications of our genetic and evolutionary unity with Nature. Grasping our ecosystemic selves, we are aware of our participation in an extensive and multi-scale life system that is here and now, minute-by-minute, second-by-second, with every molecule of O_2 we inhale and every single calorie of food energy we ingest, burn, and live by. This unity with Nature was also relevant to evolutionary dynamics far back in time; holistic life—environment relations were active when proto-humans evolved larger brain size, and when innovations such as tool use and language arose. Ecological network science reveals and quantifies that we are an incredible nexus of a multitude of energy, material, informational, and relational flows. And holistic ecology demonstrates that this network aspect is now, always has been, and always will be essential to our humanness. Humanness is essentially about connectivity to ourselves, to each other, and to the environment. We are connected whether we recognize and embrace it or not. As Wendell Berry wrote, "There is no such thing as autonomy; there is only a distinction between responsible and irresponsible dependence" (1977, p. 116). We hope this book can lead to a more responsible, engaged, and appreciative understanding of the active role of the "ecological theater" in which the "evolutionary play" is performed.

This new unity, the holistic human—environment relation, is the continuing and everyday reality, and the one we need to heed now to make a radical course correction. Ironically, understanding and appreciating our essential unity with our planetary home may well be a necessary developmental step before colonizing life beyond Earth can have a chance to succeed.

One of the surprises, delights, and rewards of the work and writing is the learning that can occur during the process. One new insight gleaned from the collaborative creation of this book involved a deeper understanding and surprising twist to one of Donella Meadows' contributions. We had known that she ranked "the power to transcend paradigms" as the number one source of leverage for influencing complex social, economic, environmental systems. At the start of this book journey, we took this as evidence and encouragement to push forward with our efforts for a paradigm shift in life sciences and wider industrial culture. As we had to struggle to compile and unify the full book, however, her principle took on new and unexpected meaning and power. For Meadows did not only speak about transcending the dominant mechanistic paradigm of modern science—her principle was generic; thus, it must apply to all paradigms. In the exhilarating but humbling epiphany that this realization unleashed, we had to apply her principle even to our own most closely held and cherished ideas and concepts. We ask that you read with an eye to assess whether we have been true to her insight. We now see even more leverage and power in being open to changing one's own mind and being free to honestly challenge one's own cherished assumptions and worldview. And we have employed her idea of "transcending" as a central concept throughout the book—one that is complementary to our other main concept, "sustaining."

We have worked in our fields for about 25 years. We have worked on this book for about 18 months. We know it is just another stepping stone. And so, as we offer this for your reading and consideration, we also ask and invite your collaboration. We seek allies willing to read and discuss this book in a graduate seminar, or reading groups, collaborative teams or interested individuals to read this closely and provide us honest feedback and criticism. We know already that a second edition will be needed, and we hope this next book will have even more coauthors and contributors.

As we continue to learn from our mentors, we also continue to deepen the process of learning from what they have learned from—natural ecosystems. While we are concerned by the state of the world, and though one key impetus is seeking to solve a systemic crisis, we are continually inspired by natural ecosystems that are able to heal themselves and to reorganize, recover, and regenerate from massive disturbance. While some of the explanations for these capacities for self-sustaining are paradoxical—requiring equal respect for life as well as death, for example—we take heart in the nearly 4-billion-year track record of success. We have lots of science to do, lots of work, and lots of creative play and fun to be had. This is the glorious bright side of the human–environment condition. In this creative space, we can solve all problems, together.

Daniel A. Fiscus
Western Maryland Food Council, Cumberland, MD, United States

Brian D. Fath
Department of Biological Sciences, Towson University, Towson, MD, United States;
Advanced Systems Analysis Program, International Institute for Applied Systems Analysis,
Laxenburg, Austria

August 2018

Reviews

Foundations for Sustainability is an important synthesis that points the way, along a long and rocky road, to an ultimate solution of humanity's environmental problem.

Edward O. Wilson, Emeritus Professor, Harvard University.

By taking the big picture of our spaceship called planet Earth, and mixing this with modern ideas on ecology and energy flows, this wonderful little book is a must read for every free thinking individual. The writing is bright, fresh and easy, and the ideas are all worth pondering for days. The Book's framework offers concrete sustainability pathways each embedded with a win-win philosophy."

Jaia Syvitski, former Director IGBP.

In clear and compelling language, the authors of this remarkable book present solid scientific and ethical foundations for a 'science in service to Life,' oriented unequivocally toward building a sustainable future. In view of the frequent co-option and distortion of the concept of sustainability, their effort could not be more timely. I recommend Foundations for Sustainability warmly to anyone concerned about the future of humanity.

Fritjof Capra, author of *The Web of Life*, coauthor of *The Systems View of Life*.

In this compelling call to "serve life", Fiscus and Fath have taken direct aim at the one thing holding us back from moving beyond industrialism: our inability to recognize the limits of our own beliefs. Drawing on an astonishingly broad and deep integration of science, philosophy, spirituality, and culture, they show us how to let go of the myths of modernity and embrace a fundamentally different relationship with each other and the planet nurtured by mutual understanding, cooperation, acceptance and unity.

Laura Lengnick, Author of *Resilient Agriculture: Cultivating Food Systems for a Changing Climate*.

Fiscus and Fath provide a comprehensive and philosophical perspective on the holistic approach to ecosystems, and on the need to confront the interplay between the physical environment and life. They provide counterpoint to those such as me, for whom the complex adaptive system perspective views system-level properties as largely emergent from processes at much lower levels of organization, but feeding back to influence those lower-level interactions.

Simon Levin, Professor in Ecology and Evolutionary Biology, Princeton University.

An ambitious work, Foundations for Sustainability *is a radical challenge to the foundations of life science. Get ready to experience your mind stretch to see the connectedness of all life in new ways. Best of all, Fiscus and Fath help us see possibilities to address our planet's greatest crises. Great for the general reader and professionals alike.*

Frances Moore Lappé, author of *Diet for a Small Planet* and *EcoMind*.

Acknowledgments

Any book project is the culmination of work made by many hands. This project, in particular, has emerged out of conversations, dialogues, collaborations, workshops, and courses that we have participated in over a lifetime of pondering big questions about the future of human–environment interactions. It is impossible to list here all the persons that have touched or shaped us and the ideas herein, but we are grateful for and anticipate further collaborations. We are thankful for the patient and responsive team at Elsevier for shepherding us through this process. We are grateful that Laura Kelleher got the whole project started and with faith and encouragement facilitated the early days of the book.

Thanks to countless librarians who help provide access to the collective wisdom of prior workers. Special thanks to Felicity Pors at the Niels Bohr Archive for assistance in tracking down the original sources of Neils Bohr's comments on "great truths".

Thanks to Sarah McManus, Steven Fiscus, and Jan Heath for generously sharing their artwork and for creative discussions of the interplay between art, science, and the main ideas of the book.

To solve a difficult problem, enlarge it

1

INTRODUCTION

At its center, this book is a scientific work that we offer in service to life itself, life as a unified whole. We have sought to develop and present scientific theory and linked applications that are rigorously rooted in science. In order to do that, we have had to question and modify the foundations of science itself, to prepare this ground so it is fertile to accept and nurture the roots of the science and actions we see as vital. The novelty and gravity of the human-environmental circumstances we now face provide the necessity as mother of inventing solutions commensurate with the challenge. These modified foundations unfold throughout the book as we aim to expand on an applied-theory science for sustainability that:

1. Balances and synergizes holism with reductionism;
2. Equally emphasizes internalist and self-referential as well as objectivist perspectives;
3. Is anticipatory and accelerates the pace and process of paradigm shifts; and
4. Is consciously, intentionally, and transparently value-based centered on the value of life.

We will elaborate on these and additional founding principles and give credit to path breaking works of those from whom we have learned. In particular, Ulanowicz' (1999, 2009) "ecological metaphysic" is a guiding light that encapsulates a system of ideas compatible with Life—environment unity. His metaphysic includes three key tenets of reality that he gleaned from studies of ecosystems and networks; living systems are characterized by contingency, feedback, and memory.

We also develop and explain the coherent links to philosophy and values as well as to actions and daily life. To address such a broad range of topics requires that we focus on a level that is shared common ground between these areas—the basic ideas at the foundation of science as well as ethics, culture, and day-to-day reality. There are many ideas and processes at the intersection, from experience and experiments to reasoning, values, decision-making, learning, and understanding. We see it as necessary to work on the bridge areas between science and society as this is where we see both the causes and solutions of our current major problems to lie.

Foundations for Sustainability. DOI: https://doi.org/10.1016/B978-0-12-811460-5.00001-7

A critical issue for understanding our current world situation—especially the "global ecological crisis," but also many related social and economic troubles—relates to the interdependence between "systems of ideas" in various cultures and subsets of cultures and "real world systems" or more simply, the real world. We will examine many detailed descriptions of the various subset problems that make up the current crisis, but for now, we cite these top 11 factors (Table 1.1) as sufficient hard evidence, corroborated by thousands of scientific studies, and reported experiences of millions of citizens worldwide, to make the case for a bona fide global socio-ecological crisis.

In addition to these primarily environmental indicators of systemic dysfunction, we could add many others that are more social and economic, such as growing income inequality, widespread armed conflicts—many of which derive from natural resources—challenges with human health, and more. Although the challenges we face are global and cannot be untangled along national borders, this critique and proposal for solutions mainly applies to industrial cultures such as the United States and other developed nations.

This is the real world as we see it now, and we aim to show that these conditions have been manifest based on the ways we think about—and then relate to—the world and environment. This interface and integral relationship between how we think and the outcomes we see in the world are at the crux of what we addressed in the 2012 paper (Fiscus et al., 2012) and other works. We proposed that our shared "system of ideas" (or paradigm, shared mental model, etc.) is responsible for our current life–threatening state of affairs in the real world showing chronic and systemic environmental degradation, as well as systemic social dysfunction. And that, going forward, in order to solve our current suite of chronic and systemic environmental problems, we will need to change our minds, mindsets, and one or more "system of ideas."

Table 1.1 Hard Evidence of Our Crisis (and see Fig. 1.1)

1. Soil loss and degradation
2. Unprecedented land use change and conversion of natural habitat to human dominated landscapes
3. Rates of species extinctions on par with the five mass extinctions of all time
4. Plateau of food production and increasing vulnerability of the food supply
5. Disruption of the global nitrogen cycle
6. Pressure related to fossil-fuel dependency (including conflicts over pipelines, fracking, and more)
7. Global climate disruption
8. Sea level rise and impacts on coastal areas with dense human population
9. Ocean acidification and related disruption to coral reefs and ecosystems
10. Water pollution and shortages in many areas
11. Persistent and bioaccumulating toxins and solid waste such as plastic and micro-plastics

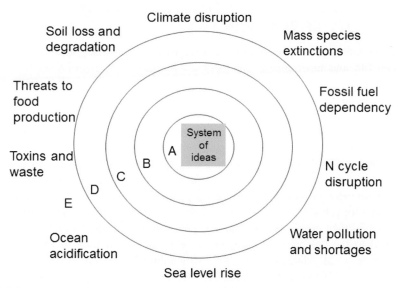

FIGURE 1.1

This diagram depicts the current human—environment system of problems or "mess" in the sense of Ackoff (1974). Our culturally shared system of ideas is at the center, and this is linked closely to the dominant scientific paradigm. This central systemic cause leads to the ultimate "tragedy of the commons" outcomes by a set of relationships that radiate outward (influence travels from center outward, from ideas to actions to global impacts) in stepwise fashion:

A. A paradigm in science is promulgated via education, technology, and media to most people in industrial culture. It is the assumed and shared normative system of ideas, specifically with respect to life and environment. In this science paradigm, life is separated from environment thus severing the unity of life and life-support systems conceptually and scientifically.

B. Following the scientific paradigm, inherent in the cultural system of ideas, life is separated from environment in mind and action. This is the key error (or outdated paradigm) that severs the unity of life and life-support systems in the real world.

C. Once fragmented, it is possible and likely that the value of environment is seen and treated as less than the value of life. Note that it is not possible for this relative devaluation of environment to occur if life—environment remains unified as a single focal entity and system of study. This follows from the revised paradigm we present here.

D. Individuals act for self-interest primarily and compete for what they perceive as limited, scarce, and zero-sum resources and assume that environment degradation is normal, expected, inevitable, and acceptable.

E. Environment is consumed and degraded as manifest in many symptoms of ecological crisis, and the influence of the citizens' mental fragmentation and devaluation of environment travels upward to larger scales and produces the global crisis.

THE REALITY OF WIN–WIN

The prevailing science paradigm sees a fragmented and antagonistic relationship between life and environment and between humans and environment by logical extension. This paradigm separates life and environment as distinct entities and is conceptually aligned with the Darwinian story of life as "the struggle for existence" emphasizing competition of individuals upon an environmental stage. Our alternative paradigm emphasizes a mutualistic relationship between life and environment (Lovelock, 1972; Patten, 1982; Fath and Patten, 1998; Bondavalli and Ulanowicz, 1999; Fath, 2007). This approach integrates life and environment into a unified whole and seeks to understand the interdependence and coevolution of the full life–environment system emphasizing cooperation. We assert that this alternative view is key for achieving a win–win relationship between humans and environment, which would then enable lasting and systemic solution (in the form of a system of solutions) to the "global ecological crisis" (for characterization of this crisis, see Wackernagel et al., 2002; Leigh, 2005; Millennium Ecosystem Assessment, 2005; Cabrera et al., 2008; Rockström et al., 2009) and environmental sustainability.

Ecological network analysis (ENA) is a holistic scientific approach that quantifies storages and flows of key life currencies such as energy, carbon, nitrogen, phosphorus, water, and more. ENA has been employed to study hundreds of ecosystems and food webs with data to document "who eats whom" and by how much in any given ecosystem of study. It is based on the science and mathematics of thermodynamics, information theory, and material flow of networks that are generic, and it has also been applied to flows of money and goods in economies, geographic flows in transportation networks, and more.

In addition to standard ecological questions of feeding interactions, ENA of food webs gives insight into the indirect relations that exist in these interconnected networks. In particular, it has been shown (Fath and Patten, 1998) that the overall relations tend toward positive outcomes both in terms of type (mutualism) and degree (synergism). In other words, normal ecosystem interdependencies promote mutualism and synergism. When all direct effects (from first order, proximate interactions) and indirect effects (from higher order interactions, between network components that are only indirectly connected) are integrated, most pairwise relations between species or components in ecosystem networks are win–win. Despite this widespread pattern of mutually beneficial relations between living entities, humans (again, in modern industrial cultures) defy this pattern and clearly show negative impacts on other species, the atmosphere, soils, other integral components of the biosphere, and even ourselves. Therefore, it is reasonable to propose our global ecological predicament this way:

> The fundamental, net human–environment relationship is antagonistic or win–lose.

The crux of the problem and solution is a shift between two scientific views of the fundamental relationship between life and environment. Thus, we propose the holistic summary of our global ecological solution, and the desired outcome we seek to assist is:

> The fundamental, net human—environment relationship is mutualistic or win—win.

With this systemic solution, we would live in a world in which human actions naturally and consistently result in *improvement* in our environment and life-support systems over time, unlike the trend of environmental degradation we see now.

A NEW SYSTEM OF IDEAS

Frances Moore Lappé presented a similar approach in her book, *EcoMind*, where her solution is that we "think like an ecosystem." Capra and Luisi, in their book, *The Systems View of Life*, suggested the solution is a systems view of life to replace a mechanical metaphor. Our recent work with additional coauthors, *Flourishing within the Limits of Growth: Following Nature's Way*, also proposed implementing lessons from systems ecology and natural processes as a step forward to find integrated, win—win solutions to the socio-ecological crisis. Here, we weave these and other contributions into a coherent framework we see as uniquely able to allow us to think, act, live, and create a culture from the locus of "life—environment as a unified whole," such that we think and act on behalf of this whole. David Orr (2008, Encyclopedia of Ecology) echoed a similar analysis as related to education:

> In important respects, all education is environmental education, that is, by what is included or excluded students are taught that they are part of or apart from ecological systems. The standard, discipline-centric curriculum may have contributed to a mindset that helped to create environmental problems by separating subjects into boxes and conceptually by separating people from nature. (p. 1119)

Many workers have used many different terms related to the shared cultural "system of ideas." It will help to acknowledge many of them and then to adopt a single terminology for this book for the sake of clarity. We are aware of this set and acknowledge many others not in this list:

Paradigm	Conceptual framework	Conceptual model
World view	Epistemology	Relational model
System of ideas	Mental model	World hypothesis (Pepper)
Mindset	Belief system	Root metaphor (Pepper)
Value system	Zeitgeist	Way of thinking
Ideology	Shared story/narrative	Metaphysic

And even others, and combinations of the above.

Given these options, we choose to use primarily the term *system of ideas* to represent the detailed and comprehensive picture and description of the shared cultural mindset; and, we use the term *root metaphor* (Pepper, 1942) to speak of the shorthand version or central motivating image to which the entire system of ideas of a given era is often distilled or condensed. We also follow Goerner et al. (2008) and their depiction of the revolution and total re-structuring process of the cultural system of ideas as *Great Change* and that the process of periodic and profound Great Change—from the Renaissance to the Industrial Revolution and more—is driven primarily by *learning*. We also use the term *paradigm* when speaking specifically of the system of ideas within science, following Kuhn (1962).

We also borrow from Goerner et al. (2008) and Pepper (1942) and see the root metaphor now in a process of changing from "machine" or "mechanism" to "ecosystem," "web," or "network." We say "now in a process of changing" and do believe qualitative and profound change is happening in the near term, but we also know that aspects of this transition have been conceived, written, signaled, foreshadowed, promoted, and built upon for decades and even centuries. Again, speaking mainly about United States and industrial culture, from Thoreau and Emerson, to Alfred Lotka and Lawrence Henderson, to Aldo Leopold, Rachel Carson, Wendell Berry, and many others in science, arts, and diverse fields— much of our message, and many of the ideas in our system of ideas have been said before. We see the current need to add our voice to this choir, to help replay and amplify those great clear voices of the past, and to call them back to the main arena for encore performances now, when we need them the most.

We see this Great Change to a new root metaphor, and the associated major overhaul of our system of ideas, to be necessary if we are to solve the current suite of entangled, chronic, and systemic ecological problems. This is our current assessment of the best way to follow Ackoff (1974) who wrote:

> English does not contain a suitable word for "system of problems." Therefore, I have had to coin one. I choose to call such a system a *mess*.

And soon after:

> The solution to a mess can seldom be obtained by independently solving each of the problems of which it is composed. (p. 21)

We similarly see the current human−environment problem as a *mess* and that to help with the Great Change to a new root metaphor and a new system of ideas will be our best and perhaps only chance to emerge from this crisis with a successful outcome and path forward to true sustainability and win−win relationships with our environment and home. Rather than solving each problem separately, a new system of ideas (in culture) linked to a new paradigm (in science) could get at the core causes from where the problems arise. See Fig. 1.2 for our depiction of this situation.

Tragedy of the commons
Humans win, environment degrades

Bounty of the commons
Humans win, environment improves

Life   Environment

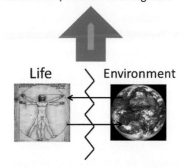

FIGURE 1.2

Contrast between the "tragedy of the commons" that occurs as a result of fragmentation of life from environment versus the possibility of the "bounty of the commons" which life achieves and which we can achieve by reuniting life and environment via foundations of science.

We need to shift from a focus on treating symptoms, something that is all too common in a highly specialized world, to preventive and systemic approaches. This is in accord with what Wendell Berry referred to as "Solving for Patterns," which relies on integrated solutions. For example, three of his fourteen principles instruct that:

> 3. A good solution improves the balances, symmetries, or harmonies within a pattern.

> 4. A good solution solves more than one problem, and it does not make new problems.

> 13. A good solution in one pattern preserves the integrity of the pattern that contains it.

> **Berry (1981)**

Those are characteristics of the solutions we seek here, and we believe our proposed solutions meet these criteria.

In this book, we present an expanded version of our previous ideas and add new evidence and explanatory corroboration. We show how and why a new mainstream system of ideas, aligned with the ecological metaphysic of Ulanowicz (1999, 2009), holds the greatest promise as a scientifically based way to better harmonize our thoughts with the real-world outcomes we desire. Speaking of this ideal end point, we could begin to describe a future human−environment system in which all of the top 11 problems above (Table 1.1) have been solved and continue in healthy and sustainable conditions. At that point, we would observe

1. Soil building, increased fertility and aggradation;
2. A compromise similar to E.O. Wilson's "Saving Half the Earth" is reached;
3. Rates of species extinctions in line with historical rates between mass extinctions;
4. Improved food production and resilience, sustainability, affordability, and health of the food supply and its beneficiaries;
5. Stable global nitrogen cycle including an end to dead zones in estuaries and gulfs;
6. Broken addiction with fossil fuels with most energy from renewable sources;
7. Stable global climate similar to the Holocene; no net increase in atmospheric gas concentrations or temperature;
8. Sea levels follow natural patterns and more secure coastal areas with human populations;
9. An end to increased ocean acidification and regeneration of coral reefs and ecosystems;
10. Abundant, affordable clean water sufficient for human needs and for wild nature; and
11. Great reduction in environmental toxins/waste; emissions within rates of recycling or decontamination.

Clearly, to achieve success on so many variables requires a new, more comprehensive, more holistic level of systems thinking coupled with well-coordinated innovative actions, monitoring, and adaptive management.

But, we do not seek to develop and propose one single system of ideas and defend it as more correct than all others in any specific absolute or exclusive sense. This kind of hegemonic approach may in fact be closely linked to our deepest underlying problem—for example, any approach that serves to divide rather than unify, splinter rather than combine views, or foster argument more than agreement seems a path toward continued fragmentation and confusion rather than unity and cooperation. Instead, we seek a "big tent" or umbrella system of ideas that is generic enough to be inclusive of and compatible with many others, even ones that are very different—even views that appear diametrically opposed. In this way, we propose a conceptual framework that we see capable to help unify people by providing a coherent and readily accessible way to see and understand our common ground.

We know of many other approaches that are compatible, and have informed our thinking, and that are also leading the way forward. Several kindred efforts that are highly noteworthy include work of the Center for Advancement of the Steady State Economy, the Great Transition Initiative, the Research Alliance for Regenerative Economics, and Future Earth. These are a few of the strong and clear voices we seek to echo and join with a larger chorus of consensus and cooperation for systemic change.

LIFE AS THE GREATEST COMMON DENOMINATOR

The need for unity instead of divisiveness, and the need for a solid basis for people to come together to address our profoundly problematic situation, can be advanced by seeing life as the "greatest common denominator." In this book, we follow Albert Schweitzer and his work on "reverence for life," and we ground our scientific and conceptual work with a view in which life is the basis of value and ethics. We see science and the scientific method as of fundamental importance and utility, but many people have become skeptical and lack in trust and respect for science. It may be that more often and more explicitly stating and ensuring that science acts in service to life will aid both the progress toward sustainability and increased trust and respect for science. These two benefits are why we see the need to begin the foundations for sustainability with ethics and values and to focus our ethics and values on the unique and uniquely unifying reality of life. Chapter 2, is devoted to this idea of "life value," but we mention Schweitzer's concept here briefly. Schweitzer (1965) wrote:

> The elemental fact, present in our consciousness every moment of our existence, is: I am life that wills to live, in the midst of life that wills to live. The mysterious fact of my will to live is that I feel a mandate to behave with sympathetic concern toward all the wills to live which exist side by side with my own. The essence of Goodness is: Preserve life, promote life, help life to achieve its highest destiny. The essence of Evil is: Destroy life, harm life, hamper the development of life.
>
> The fundamental principle of ethics, then, is reverence for life. (p. 26)

The importance of this perspective is that life itself has value, inherent to it as life. This is not a measure of the economic transactions one extolls from nature nor is it the ecological rationalization that life supports further life processes, nor the scientific value we earn through investigations, nor even the aesthetic wonder and beauty we experience from a serene sunset or breaking wave. Rather, it is a moral underpinning of value for life simply because it is living. Schweitzer also wrote:

> I cannot but have reverence for all that is called life. I cannot avoid compassion for everything that is called life. That is the beginning and the foundation of morality. (pp. 115–116)

Solutions that improve the conditions of life are sought as foundational, but we also speak soon (just below) of the distinction between two aspects of life: discrete and sustained. Clearly, promoting sustained life is a priority from the perspective of the long term and all peoples. In addition to Schweitzer, Aldo Leopold (1949) provides a succinct description of the approach to ground all ethics in explicit valuation of life:

> A thing is right when it tends to preserve the integrity, stability, and beauty of the biotic community. It is wrong when it tends otherwise. The Land Ethic, *A Sand County Almanac.*

It is interesting to note that Schweitzer's first printed mention of reverence for life came after an experience with hippos in Africa, and Leopold's views were transformed by witnessing a wolf dying, watching the green fire in its eyes extinguishing. For these scientists and thinkers, as for ourselves, close contact with life can spark lasting insight.

That life serves as an excellent universal basis for ethics, and as an ethical basis able to unify all people, is perhaps common sense. We add details below, but by speaking of "life," we refer to life itself, life as a unified whole, all life; this must be understood as distinct from the life of a single individual or organism and even as distinct from the existence versus extinction of a single species. To state the rationale explicitly:

> Life is the "thing" (quality, reality or process) we humans all share, it is the also the "thing" that unites the human species with all other species on Earth. Life is the prerequisite that must continue for any other human endeavors, concerns, hopes, aspirations, development, or progress to be possible. If life ends or is jeopardized, then all other concerns end or must take lower priority.

Additional corroboration comes when we ask what threatens life and therefore is bad, wrong, or undesirable. Here, we can reverse Leopold's statement:

> A thing is wrong when it tends to threaten or harm the integrity, stability, and beauty of the biotic community.

Note, that for Leopold, the fundamental unit of which he speaks is the community. This is an interesting and important articulation, and it is also worth noting that life appears here as an adjective—a modifier describing a certain type of community, a biotic one.

LIFE AS COMPLEX AND DUAL MODELED

We have spoken of life as a unified whole, and of the distinction of discrete versus sustained life, and this helps to understand the importance of the associations of Leopold, Berry, and Schweitzer with life as community or a complex, self-reinforcing, awesome, and eternal phenomenon. This dual-model view of life draws clear lines between the discrete life inherent in cells, organisms, and even species, and the sustained life inherent in communities, ecosystems, and the biosphere. In addition to the difference in the clarity of spatially boundedness of discrete life forms (again, mainly cells, and organisms) which have readily recognizable skin or membrane boundaries, and the difference between the longevity of their existence, we can focus on a difference related to logic. At any point in time, discrete life forms can be determined to be either alive or dead. In contrast, sustained life forms (most clearly as applied to ecosystems and the biosphere) cannot be thus classified and instead are simultaneously both alive and dead; they contain and depend on the integration of both living and nonliving functional components.

In this book, we employ and further develop these complementary models of sustained and discrete life, and we point out ways in which over-emphasis on the discrete model in life science and education is linked to our systemic life—environment crisis. Imagine if we taught physics as restricted to either the particle or the wave view of light. We see the complementarity of life models just as essential for scientific understanding and all applications.

At least three corroborating views exist for the idea of two types of life, for the necessity of the distinction, and for individual definitions for "discrete life" and "sustained life." We present just one of these for now. Morowitz (1992: 5) wrote, ". . .we recognize two approaches to defining life: one focuses on the properties of individual organisms and the other is much more global and ecological in character," and:

> Sustained life is a property of an ecological system rather than a single organism or species. A one-species ecological system is never found. The carbon cycle requires at least one primary producer and a method of returning carbon to the CO_2 pool. A system of only herbivores would eventually die of starvation. A system of only primary producers would grind to a halt from CO_2 exhaustion unless autolysis or burning produced CO_2 at a sufficient rate, which does not appear to occur.
>
> **Morowitz (1992, p. 54)**

He went on to discuss how the origin of life can be considered to involve key planetary processes including "protoecological cycles" and synergy between anabolism and catabolism. Unfortunately, he may have diluted the power of his holistic framework when he wrote, in the same book, "All life is cellular in nature," and later, "A cell is the most elementary unit that can sustain life." Perhaps, by not keeping the terminology and reference units related to sustaining life clear, and by other statements such as titling his book *The Beginnings of Cellular Life*, Morowitz (1992) assisted with continuation of the single-type, species-centered paradigm of life.

More discussion and supporting work on this complementary model of discrete and sustained life is presented in Chapter 6. See also the section below on a related dialectical distinction between deep truth and simple truth. Much as we must reconcile seeming opposites and mutually exclusive models to understand the complexity of light, to understand the complexity of life, we have to integrate death and dying, eating and recycling. All discrete life forms, outside of autotrophs, kill and eat other discrete life forms to sustain their own lives, and life itself is a continual process of killing, eating, dying, going extinct in individual/ organism/species form to sustain life as a unified whole. Patten (2014, 2016a,b) has asked "Is life's destruction of current life (eating and being eaten) to sustain and re-configure new life to fit a changing planet the only way to organize a biosphere?" We currently see no way to answer other than to say "yes;" and thus, we propose to act to make the best of this reality. As for Schweitzer, for us this mystery and paradox is cause for reverence and awe. We also see this holistic integration of life and death as necessary for the path forward to a truly sustainable life—environment relation.

DYNAMIC RANGE OF APPLICABILITY

We present in this book ideas able to unify life and environment, to integrate discrete and sustained forms of life, and to reconcile that sustained life depends fully on death. In Berry's (1981) words:

> In an energy economy appropriate to the use of biological energy, all bodies, plant and animal and human, are joined in a kind of energy community. ... They are indissolubly linked in complex patterns of energy exchange. They die into each other's life, live into each other's death. (p. 90)

One depiction of the "dynamic range" of this holistic conceptual framework is shown in Table 1.2—a chart of six realms. This table shows our ambitious hope for a system of ideas to link from ideas of, and beliefs about, God, spirit, and the unknowable on the left side to real world and practical action for daily life, meeting real needs of the human body, and community action for human and environmental enhancement on the right side. One need not believe in God for this framework to be of value—one could easily start with philosophy, ethics, and values and consider just five realms.

Table 1.2 represents a conceptual framework toward a system of ideas able to span from the unknowable and mysterious, to the most abstract and subjective ideas, and on to the most pragmatic and basic aspects of daily life with the ultimate goal of healthy and sustainable human—environment relations. The main contributions of this book are in the center of this table, and they are central to the new system of ideas we see now emerging. Near the middle of Table 1.2, we see our ideas as pivotal, serving catalyst roles, and providing leverage for systemic change—including a new science formalism that prohibits fragmentation of life from environment and of discrete from sustained life, and Ecological Network Analysis showing how "everything is connected to everything" very literally and how this can be quantified (Ulanowicz and Puccia, 1990; Fath, 2007).

In Chapter 6, we show in detail many more major and paradigm-shifting concepts from ENA and whole system studies. For example, such studies have shown that the US human food system uses 10 calories of fossil energy for every 1 calorie food energy, which is clearly unsustainable, and clearly different from energy use in natural food webs. ENA also quantifies how indirect effects between any two participants in a food web are usually greater than direct effects. The science of ENA is the most versatile tool by which observations, data, and replicable science can be part of the Great Change from the mechanism to the web root metaphor.

Table 1.2 Dynamic Range of Applicability—The Ambitious Goal of Our System of Ideas, Unified via Holism, Reverence for Life, and Love

Realm of Pure Spirit	Realm of Human Spirit	Ethics, Values	Science Theory	Science Application	Society/Enviro. Realm
Higher power	Love thy neighbor God is Love Golden Rule	Service to…??? (To life, to God, to nation, to humanity, to community, to family, etc.)	Holistic Science to Serve Life Itself, Life as a Unified Whole		Systemic Change for Health and Justice (some specific and personal examples below)
God	Love	Love	Love	Love	Love
Unknowable	Faith/belief ← Link to unknowable	Ideals Schweitzer: Reverence For Life	Potential Anticipation	Actual Reality	Living reality Gratitude
Beyond our ken	Reflection Wisdom traditions Religions Truth, meaning	What is right to do? What is important?	Creation, ideation Holism How to observe What to look for	Validated data Ex. Networks Ex. Human food web Care, compassion	Sustainable agriculture Food assistance Anti-poverty programs
	One spirit/soul	One heart	One mind	One body	One body
	Spirituality Spiritual practice Meditation	Philosophy Ethics Values	Science as theory Ideas Mental models	Science as applied Data Maps and graphs	Service to people Community action
Note: missing from this table: war, greed, money, fear, hate…	Church/Temple/Mosque/Synagogue/Shrine/etc.	Subjectivity	More objective, hypotheses	Observing process, real observations	Environmental stewardship
Life	Life	Life	Life	Life	Life

Holistic spectrum of spirit-material

Life (again, life itself, or life as a complex unified whole) is such a comprehensive basis for ethics and unity that it can even serve to unify and integrate two widely divergent human views. Next, we present an example showing how we seek to aid unity between people and groups with divergent views and belief systems—that is, operating with different systems of ideas—related to the environment.

SUSTAINERS AND TRANSCENDERS

One may get the sense that we move one step forward and two steps back, that the right hand doesn't know what the left hand is doing, and that we have major institutions, policies, movements, energies, leaders, organizations, and efforts working at cross purposes with respect to sustainability and related challenges in the human—environment crisis realm. Another sense that has stimulated this work comes from efforts to understand the "climate deniers" and similar groups who are so opposed to the mainstream scientific assessment of climate change and the warnings that major change is needed to avert or reduce catastrophic impacts. Unfortunately, climate has become one issue in the bundle of positions that right and left politicians carry, particularly in the United States. There is no doubt that some positions are driven by financial interests, but at the core of the divide may be an oppositional base that cannot reconcile the climate issue with their inherent, adopted, learned, assumed, or operating system of ideas. Problems occur, however for both sides, when a system of ideas mismatches to reality, as well as when opposing sides fight and impede each other to the detriment of both. However, what is that system of ideas, or root metaphor, that sees no urgency of priority to limits and planetary boundaries, or that acts to encourage individual views that are diametrically opposed?

We think it will help to characterize two tendencies or modes of life that play out in human systems. We suggest that if we do not understand these two modes we may continue what we have now—huge amounts of unnecessary confusion and conflict. These ideas came out of trying to understand how someone could have a system of ideas (again, we could say mental model or belief system) such that they do not accept the scientific consensus of climate change or evolution. We struggled to find a larger explanation (in the spirit of one of Stephen Covey's *Seven Habits of Highly Effective People* that is "seek first to understand and second to be understood") such that perhaps the "climate deniers" were "right" on some other level that we were not aware of and not taking into account. We express our simple working hypothesis with metaphors of two types of people, with two different systems of ideas, with respect to environment and especially with respect to environmental limits, constraints, or challenges.

This typology starts by assuming that everyone is aware of the environmental limits, constraints, and challenges now increasingly apparent, at least at some general level. However, the proposal is to see two different responses to this awareness. These two groups we label as "Sustainers" and "Transcenders," in order to make the typology value-neutral similar to Myers–Briggs personality types. We are not trying to say that either is "good" or "bad," or that one is better than the other in an absolute sense.

In this typology, the "Transcenders" are aware of environmental limitations (again, even if just at some general level such as "human impacts are damaging the environment," or "we are starting to see real shortages of key natural resources"), but when confronted with a perceived limitation they seek to break through it, innovate out of it, or change the world to **transcend** that external barrier or limitation. The "Sustainers" adopt the opposite approach, and they accept the perceived environmental limit and then seek to change themselves to fit within, and **sustain** a lifestyle and culture within, that perceived real world constraint.

Based on holistic and historical understanding of the origin, evolution, and development of life, as well as understanding of human origins, evolution, and development, we could view these two tendencies as both natural and needed. The Transcender may be first out of the block, as in the case of migration to new areas, modifying an environment to satisfy essential human needs. Whereas, this approach focuses on extensive variables (quantities and acquisition), the Sustainer contributes to resource intensivities (qualities and management). This can be seen akin to the pioneer and climax adapted species described in ecological succession theory. The early arrivers are attuned for growth, rapid response, and facilitating the environmental conditions (r-selected species), whereas the later ones are prepared to persist over the long haul with a priority on stability and homeostasis at or near the carrying capacity (K-selected species). Since there is natural variety over space and time, both roles contribute to the overall functioning in a patchy landscape.

If these two modes are both natural and needed, if they have existed forever, and are likely to exist forever, then we should not be stuck in a battle in which adherents of the two worldviews fight against each other. It would be better to dispel any myths or assumptions that either of these modes is more right or absolute in truth. At a local scale, there may be a case of "right time, right place" that one worldview fits the situational conditions "better" when resources are plentiful (Transcenders) or scarce (Sustainers), but in a diverse and multiscale system there are typically niches satisfying and requiring both types.

Elbow (1986) provides a generic principle of cognition that is relevant here:

> The dialectical pattern of thinking provides some relief from [the] difficulty inherent in knowing. Since perception and cognition are processes in which the organism "constructs" what it sees or thinks according to models already there, the organism tends to throw away or distort material that does not fit this model. The surest way to get hold of what your present frame blinds you to is to try to adopt the opposite frame, that is, to reverse your model. A person who can live with contradiction and exploit it—who can use conflicting models—can simply see and think *more*. (p. 241)

We examine and employ Elbow's ideas in greater depth in Chapter 7, but we see this concept as important at the beginning. If we could agree that our models—as good, true, and useful as they are—also *blind* us to any aspects of reality that are not included in the model, then this highly refined and mature level of dialectical thinking could be of enormous value in ending petty conflict so that we can cooperate to tackle the real-world challenges we must face together. This blindness becomes increasingly evident in the later chapters dealing with the dominant machine metaphor in science and its applications.

One futuristic example of a case in which both types are needed is any project to colonize life beyond Earth. As remote a project as it may sound, several groups are actively planning and designing such missions now, such as the plans of SpaceX to colonize Mars. For this project (which is fraught with many issues, controversies, and unknowns, but we think still serves well as a case study and thought experiment), we have to have the two camps—the Transcenders and the Sustainers—working in close collaboration and synergy. Individuals from each camp must work with acceptance of each other despite their seeming diametric opposition. The Transcenders might lead the way, since the mission is perhaps primarily about transcending the limit of the Earth as a base for life and human life. But the Sustainers must co-lead, since for the mission to succeed in full—an independent and self-sustaining outpost for life and human life beyond Earth—we not only have to leave Earth, but we have to know how to take with us the crucial, holistic, resilient, complex life-support system all humans must have. We have to take with us a mini-biosphere, and we have to be able to self-sustain in space as we colonize. This would be true for very long distance and long-term space travel, for long-term life on a space station, and for establishing and sustaining life on any other planet. NASA and other space efforts so far have had to work to understand how, and then to develop successful systems, to recycle the biological and other wastes of astronauts—CO_2, water, urine, feces, food wastes, and more. But as we imagine and begin to design the qualitatively different missions of true colonization—missions which must be designed such that the crafts, crews, and systems are fully independent of Earth—at that point we must understand how, and then to develop successful systems, to recycle astronauts themselves. We will have to understand the life−environment relation at its very root and complex dialectical core in order to establish a successful, survivable life−environment relation in the context of some totally different space or planetary environment.

A space colony mission needs the full unity and cooperation of Transcenders and Sustainers due to lack of contact with Earth. And, we need both modes cooperating on Earth too, since we only have one Earth. Thus, figuring this out, ending the conflict and confusion on Earth, is a top priority. Talk of space is important in its own right, but also serves for discussion purposes and to examine the broad and perhaps universal applicability of the typology. Remember—Earth is very much like a spaceship (Spaceship Earth of Kenneth Boulding, Buckminster Fuller, and others) and very much a solo life colony.

This Sustainer—Transcender typology relates to others already published and developed, including various typologies of people and their worldviews, systems of ideas, and root metaphors with respect to the environment. We consider just one here in detail but see value in exploring other such type schemes to aid with unifying and finding synergy. In his scheme with "Druids" and "Engineers," for example, Paul Saffo seeks to prevent a viewpoint divide that could "…frustrate the sensible application of technological innovation in the service of solving humankind's greatest challenges." (Saffo, 2013).

Thompson et al. (1990) proposed a theory of culture and characterized a typology of several distinct "ways of life" that provides support for the typology proposed here. They base their work on the grid-group typology of Mary Douglas.

Their work employs three key concepts and terms: cultural biases, social relations, and ways of life. They unify these three: "When we wish to designate a viable combination of social relations and cultural bias, we speak of a *way of life*." Using these concepts, they "present a theory of sociocultural viability that explains how ways of life maintain (and fail to maintain) themselves." (Thompson et al., 1990, p. 1). They wrote:

> Causal priority, in our perception of ways of life, is given neither to cultural bias nor to social relations. Rather each is essential to the other. Relations and biases are reciprocal, interacting and mutually reinforcing: Adherence to a certain pattern of social relationships generates a distinctive way of looking at the world; adherence to a certain worldview legitimizes a corresponding type of social relations. As in the case of the chicken and the egg, it is sufficient to show that cultural biases and social relations are responsible for one another, without confronting the issue of which came first. (p. 1)

In particular, the cultural theory stereotypes are useful in understanding one's relation with the environment and whether the focus is on managing and changing ourselves, the environment, both, or neither. Table 1.3 shows the combinations of interactions.

Our Transcenders and Sustainers map onto the Individualists and Egalitarians, respectively. The centrist, compromising Hierarchists represent a pragmatic middle ground and the capricious Fatalists just try to get through each day. This can also be represented by the now classic ball and cup stability diagrams (see Fig. 1.3) indicating whether a system, represented by the ball, is stable or not in its current landscape. The Egalitarian frame is precarious with any disturbance

Table 1.3 Cultural Theory Types of Thompson et al. (1990) with Our Sustainers and Transcenders Shown

	Manage Needs	**Not Manage Needs**
Manage resources	Hierarchist	Individualist (Transcenders)
Not manage resources	Egalitarian (Sustainers)	Fatalist

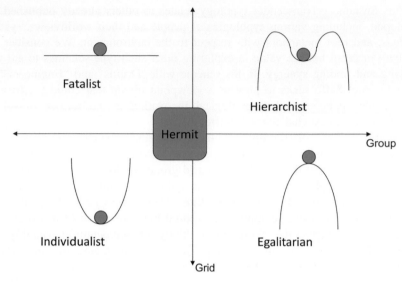

FIGURE 1.3

Ball and cup diagrams depicting systems and stabilities from the Thompson et al. (1990) worldviews. This includes the center group hermits which can show characteristics of any group.

throwing it out of balance and into a new regime (pick a daisy, move a star). Oppositely, the Individualist sees a world of deep stability that any action we take will be absorbed and buffered by the forces of nature (how can one species, humans, impact the vast atmosphere?). Hierarchists again play the role of compromiser recognizing both buffers and limits. Poor Fatalists see no hope for stability in their flatworld where anything goes. This typology provides a useful guidepost to understand why one holds specific positions; however, it is important to note that these labels are not absolute and can and do change under different circumstances and situations. A person making decisions on one topic as an individualist could be a hierarchist on another (Fig. 1.4).

In summary, our main purpose in defining Transcenders and Sustainers is toward seeking reconciliation and cooperation. Rather than continue to fight, we need ways for these two strongly opinionated camps to accept and respect each other for the greater good. It may help Sustainers to consider that perhaps it is not so much that Transcenders—or, as sometimes disparagingly called "climate deniers"—literally argue or refute evidence of environmental limits like climate change, it may be more that they see the whole topic as an irrelevant or "unthinkable" idea or issue. Similarly, rather than convert Sustainers to a more growth-oriented view and action plan, Transcenders may do better to see the widespread need for learning to manage human needs and working with rather than against the inherent patterns and processes of nature that have helped life to self-sustain for over 4 billion years.

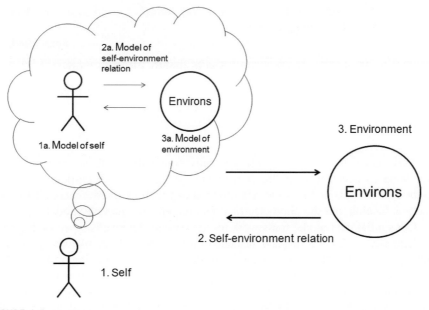

FIGURE 1.4

Depiction of self-world model in mind. This cartoon illustrates a simple idea—two people can share the same real-world environment but have very different models in mind by which they perceive and understand that shared objective reality. Different models or systems of ideas can also lead to different expectations about the best or normal human—environment relation.

DEEP TRUTHS AND PARADIGMS

Our starting point is to develop and present ideas, theory, science, data analysis, applications, and examples that can help us achieve true environmental sustainability. However, given that this ultimate goal is so completely entangled with human social, economic, and environmental ideas, policies, systems, and practices, and with ideologies, beliefs, and governance strategies, we see it as an integral goal to help achieve human peace, justice, and unity, and overall well-being, thus requiring widespread cooperation and greatly diminished conflict, confusion, divisiveness, and disagreement. This grand goal is again our approach to "start with the end in mind" and "to think win—win". This also acknowledges that people are very different and diversity is good.

One approach to unify at deeper levels what seem to be in conflict on the surface is the idea of "deep truth" from Niels Bohr. Bohr spoke of a deep truth as a truth for which its opposite is also a deep truth. He wrote:

> ...the old saying of the two kinds of truth. To the one kind belong statements so simple and clear that the opposite assertion obviously could not be

defended. The other kind, the so-called "deep truths", are statements in which the opposite also contains deep truth.

Bohr (1949)

His son, Hans Bohr, wrote of this view in slightly different terms:

One of the favorite maxims of my father was the distinction between the two sorts of truths, profound truths recognized by the fact that the opposite is also a profound truth, in contrast to trivialities where opposites are obviously absurd.

Rosental (1967)

This openness to opposing ideas is fully compatible with science, as in the simultaneous acceptance of the dual models of light as both particle and wave. Light can be explained as wave via observations such as interference, diffraction, and polarization, but the photoelectric effect requires the particle model.

Our similar dual-model framework for life presented above is based on the complementarity of discrete life and sustained life, and we believe likewise presents a fuller picture than either taken alone. If this discussion of reconciling opposing ideas seems very abstract and that we have strayed from our assertion of real-world relevance, then consider Donella Meadows' ranked list of sources of leverage in complex systems. Meadows (1999) was a systems thinker and lead-author of the landmark book, *Limits to Growth*, among many lasting contributions. She understood the profound complexity of real-world systems, and the difficulties of modeling them, but she also was highly disciplined, wise, and clear about effective ways to engage, intervene in and even "dance with" (Meadows, 2001) complex systems so as to be successful in steering them, and ourselves, toward better futures.

Meadows (1999) proposed 12 leverage points for change with complex socio-environmental systems, which she referred to as "points of power" since a small change in one can produce a large effect on the system. For our purposes of unifying people of seemingly opposite perspectives, note especially the difference between her #1 and #2 sources of leverage (Table 1.4).

This article is a mother lode of important concepts, but we quote just one about her penultimate leverage point #2: "...people who have managed to intervene in systems at the level of paradigm have hit a leverage point that totally transforms systems." (Meadows, 1999, p. 20). She also explains that societies resist paradigm change ruthlessly, while it can occur instantaneously for an individual, much as for Leopold's land ethic and Schweitzer's reverence for life epiphanies. Given our need for transformative, systemic, widespread societal change, we see it as mandatory to explore this leverage point fully.

We had not really understood the profound relevance of her #1 source of leverage—the power to transcend paradigms—until after an extended period of time considering what it would mean not only to let go of our own cherished idea that the Sustainer paradigm is more correct, more true, and more good, but also to accept that its total opposite, the Transcender paradigm, is equally true and good. This is akin to Meadows' #1 point of leverage, in the sense that to

Table 1.4 Meadows' Leverage Points

Places to intervene in a system (in increasing order of effectiveness)

12. Constants, parameters, numbers (such as subsidies, taxes, and standards)

11. The sizes of buffers and other stabilizing stocks, relative to their flows

10. The structure of material stocks and flows (such as transport networks, population age structures)

9. The lengths of delays, relative to the rate of system change

8. The strength of negative feedback loops, relative to the impacts they are trying to correct against

7. The gain around driving positive feedback loops

6. The structure of information flows (who does and does not have access to what kinds of information)

5. The rules of the system (such as incentives, punishments, constraints)

4. The power to add, change, evolve, or self-organize system structure

3. The goals of the system

2. The mindset or paradigm out of which the system—its goals, structure, rules, delays, parameters—arises

1. The power to transcend paradigms

transcend paradigms is related to the ability to let go of attachment to or defense of one's own most closely held point of view. It feels as if a much larger and broader form of unity or conflict resolution is possible when we go all the way beyond specific paradigms as Meadows suggested. While some sort of "victory" of Sustainers over Transcenders might help to resolve our current planetary ecological crisis in the shorter term, mutual understanding, acceptance, unity, and cooperation between both would seem to promise success of human life, and life itself, even beyond planet Earth and over the much longer term.

Of this greatest source of leverage, Meadows (1999) wrote: "It is in this space of mastery over paradigms that people throw off addictions, live in constant joy, bring down empires, found religions, get locked up or 'disappeared' or shot, and have impacts that last for millennia." She concludes: "In the end, it seems that power has less to do with pushing leverage points than it does with strategically, profoundly, madly letting go." Harkening back to Thompson's et al. (1990) cultural theory, the Hermit, who is best equipped to throw off paradigms, is most needed during times of transitions or Great Changes.

We work from the scientific premise and present evidence and data along with logical, rational arguments that such a strategy shows the greatest promise to work in the real world. Ethical and spiritual views provide another means of support for this open-minded view—from the Buddhist principle to strive for nonattachment to the Christian precept to "love your enemies" both speak of opposites and letting go.

The bottom line is that life supersedes all other values because if we cannot continue life, then nothing else matters. It is perhaps ironic that pushing the life-support systems of the planet to near breaking can serve to bring into greater focus what is most important. A crisis can lead to a moment of clarity unseen in the daily routines. We are now beyond the limits of the possible, and so we are able to determine this via the rare benefit to see the limit and boundary from both sides.

The Egalitarians may want to battle the Individualists today; and, the Sustainers may want to fight the Transcenders tomorrow, but such battles, quarrels, debates, and games are either meaningless or suicidal unless we are able to continue life itself indefinitely. Thus, we also see that we must do two contradictory things—we must sustain life on Earth and we must colonize life beyond Earth (at least before Earth is destroyed, or the Sun burns out), or again: all of our cherished arguments, platforms, ideologies, and beliefs become meaningless and will terminate for all time. In such discussions of the ultimate—ultimate ends, ultimate values, ultimate priorities—we could well take things to a logical extreme conclusion: our choice appears to be whether we prefer eternal life or eternal death. If this stripped-down ultimate seems too extreme, then consider this phrasing:

> Our ultimate choice appears to be whether we commit to do everything in our power to sustain life indefinitely and thus accept no other possible future, or to allow ourselves to believe that life may end, and that may be acceptable, unavoidable, or even necessary.

This ultimate choice brings to light another cherished paradigm or system of ideas that may be linked to the root cause of our situation—the elevation of the second Law of Thermodynamics to being the supreme law of physics. While many treat entropy and the tendency to decay and disorder as the dominant tendency, in this book we describe a view of entropy as complementary with its opposite—"syntropy" or self-organization. We mention it now, and discuss it in-depth later, as another scientific concept at the foundation of our system of ideas depicting a unified life—environment system.

The great potential is that if we can reconcile the many apparent contradictions we have examined and again following Covey's Seven Habits, seek win—win and find the synergy, then we could stop fighting and start finding common ground. We—all people, people of the United States and industrial cultures—need to end conflict and start with healing, increasing our maturity, peace, wisdom, better progress toward meeting all basic human needs, ending the chronic and systemic environmental crisis, ending poverty, war, and other unnecessary ills that we could remedy if we all work together. As a step in this direction, we (authors) start with an effort to help heal this massive split as we see it as now a massive waste of time, energy, brain power and resources that we desperately need to redirect and channel in to positive solutions—even if the positive applied solutions are agreed and resources shared between priorities advocated by the "Transcender" and "Sustainer" camps.

As we present our system of ideas for understanding the life—environment relation, we will often employ "dialectical thinking" as used above to understand

Transcenders and Sustainers. We will explore modes of thinking, logic and general cognitive and conceptual strategies for conflict resolution, resolution of paradox, holistic ways to address chronic/systemic problems, and similar approaches we need to understand and solve our current human—environment crisis.

For the rest of the book the major sections and concepts will be as follows:

Chapter 2. Life as the basis of value—following Schweitzer, Leopold, and others, we go further into the effort to restate, clarify, and corroborate these ethical tenets as key motivational impulse for the whole new system of ideas and new root metaphor. A system of ideas in which the world is seen as living, and with generally useful models like system, web, network, and others. We contrast this with the view of world as dead and a root metaphor, general model of explanation centered on the idea of "mechanism." We examine Ulanowicz' work on an ecological metaphysic and a cosmology of hope.

Chapters 3—5. Holistic science to balance reductionism, repair fragmentation and unify fields splintered by hyper-specialization. The first step on this holistic approach is to see life—environment as a single unified system, entity, and relationship. We develop the framing that fragmentation causes the systemic crisis we now experience, and holism is the solution. We will contrast win—win versus win—lose life—environment relations and examine how life achieves the win—win outcome such that the environment improves (oxygen atmosphere, ozone layer, soils, etc.) but industrial human systems manifest win—lose outcomes (such as the top 11 symptoms of global environmental crisis).

Chapter 6. Seven Life Lessons we have learned from studies of ENA and systems ecology. ENA provides many holistic and systemic results, for example, the matrix of total impacts that shows "everything is connected to everything" very literally and quantifiably, not just as a loose metaphor or romantic notion. This chapter includes discussion of "coupled complementary processes," which build on the "coupled transformers" of Lotka (1925). This section also presents a novel science formalism that serves to prohibit fragmentation of the life—environment unity. This formalism employs hypersets, a self-referential logical and mathematical construct.

Chapter 7. Characterization of the two-way bridge between science and technological applications. Starting with Robert Rosen's modeling relation, we examine the profound impacts and power of the machine metaphor for transforming the world.

Chapter 8. Technology, applications, and policy based on holistic life science. We describe several case studies that fit our proposed foundations for science and recommendations for actions based on this science. These examples show how a new root metaphor based on life instead of machine can do better.

Chapter 9. From the "hard," technical and abstract realms of theory and science, we transition to everyday life. We envision thought and action from the locus of, and in the service of, life itself. We borrow from Lappé who advised that we learn to "think like an ecosystem." We end with some final discussions and qualifications, including suggesting next steps. This includes a recommendation that academic and other scientific enterprises lead by example and transform themselves to sustainable operations first.

This book focuses on the science of sustainability, but we work from the perspective to always keep in mind the opposite/complement of this Sustainer worldview and to be aware of the necessity of both Sustainers and Transcenders working in complex, dialectical, synergistic cooperation. We address these broad efforts, questions, and goals throughout this book.

1. What is a foundation for science and society that can lead to and help sustain a healthy human—environment relationship?
2. What would the world under this new science paradigm and societal system of ideas look like? What would it be like if and when we achieve a win—win human—environment relation?
3. Given a clear vision of this ideal end point, how can we get there? What combination of a new science paradigm, a new cultural shared system of ideas, a new value basis, plus then operationalizing these values and concepts, will work?

   This end point is described well in our prior book, *Flourishing Within Limits to Growth*, in which we added specific actions, policies, applications, and example success stories to the general recommendation to "follow nature's way." This present book is not mainly about the applications; this book is about the foundations that could lead to more rapid and more successful applications and aid justifying and funding for them. In fact, intentionally, in the previous book, we chose actions that are instrumental and plausible in our current governments (increased recycling, foreign aid for family planning, increased education and research, particularly in STEM fields, implementation of Pigovian taxes). These fixes did not require a paradigm shift.
4. Our prior book was primarily from and for the perspective of the Sustainers, and so in this book, we ask: How to reconcile and synergize the seemingly opposite, seemingly contradictory worldviews and values of the Transcenders and Sustainers so each aids the other rather continuing the mortal battle which now thwarts both? This topic overlaps with both what the world will look like under a new paradigm and how to get there.
5. What are the main characteristics and foundational values of other worldviews, paradigms, and systems of ideas that we expect to lead to negative futures and to fail at the core mission to value life?

We end this introduction with a quote from Wendell Berry. Berry describes well how our system of ideas and culture can arise naturally from life itself. These words, and this book, can be interpreted as a call to serve life, an elaborate and sincere invitation with many suggested methods for how to go about this great service mission.

> The concept of country, homeland, dwelling place becomes simplified as 'the environment'—that is, what surrounds us. Once we see our place, our part of the world, as *surrounding* us, we have already made a profound division between it and ourselves. We have given up the understanding—dropped it out of our language and so out of our thought—that we and our country create one another, depend on one another, are literally part of one another; that our land passes in and out of our bodies just as our bodies pass in and out of our land; that as we and our land are part of one another, so all who are living as neighbors here, human and plant and animal, are part of one another, and so cannot possibly flourish alone; that, therefore, our culture must be our response to our place, our culture and our place are images of each other and inseparably from each other, and so neither can be better than the other.
>
> **Berry (1977, p. 24).**

Life as the basis of value

2

INTRODUCTION

In Chapter 1, "To Solve a Difficult Problem, Enlarge It", we mentioned the need for a systems worldview grounded in ideas of life as the basis of value. In this chapter, we describe a set of principles, and related goals, mission, and ultimate purpose, for a new science that serves life and humanity. The principles we propose are for a value system in which the primary basis of value is life. This puts life in the dual role of both means and ends. In a holistic sense, there is closure of activity such that an individual's action makes a life and is a life. This approach is observed in the tightly coupled networks in nature and applied here to humans in both science and society.

In sustainable, holistic, and organic agriculture, this expression is often used to summarize the strategy of humans interacting with a complex living system to grow food:

> Feed the soil, and let the soil feed the plants.

Continuing our analogy that the foundations for science we propose are like fertile soil in which a new holistic science of sustainability can take root, grow, and flourish, we could say our strategy in this chapter is to:

> Feed the foundations, and let the foundations feed the science.

This is the positive and proactive principle that leads to both fruitful food production and healthy soils in the agricultural case and, we propose, can lead to both fruitful scientific advances and healthy integration of the ultimate human priorities in our science foundation's case. And, to examine the alternative helps to prove the necessity of this holistic approach—industrial agriculture that shortcuts the above strategy and works to feed plants directly by ignoring soil health and function and applying inorganic and synthetic fertilizers is known to degrade soil quality and fertility over time. This adds fragility and dependencies into the production system that are not necessary in healthy systems. In particular, this mechanized, linear, industrial approach breaks the unifying cycles of nature that include fertility, growth, *and renewal*. The soil is made vital again by the actions of decay and the role of decomposers living and making. The conscious removal of this step through biocides that render soil nearly inert displays an arrogance in our ability to control the system and manage it better than nature. This mistake of failing to care for soils, whether through modern chemicals and energies or more ancient soil/water mismanagement, is known to have resulted in the ends of several great civilizations; the lessons of history of the fundamental value of soil are clear. Similarly, we believe that our current mainstream science may be seen to shortcut the above strategy of transforming and renewing the essential role of science foundation based in the most widely shared human values. And, in the analogous quest for short-term science gains—numbers of publications, numbers of citations, grant dollars awarded, etc.—we assert that something of much greater value, essential to long-term survival of our civilization, is being missed and lost.

One first step is to clarify and set our terminology for the central concepts. In Chapter 1, and in our dual-modeled view of life, we used "life" to mean life itself, life as a unified whole, all life on Earth as integrated with environmental life support, a complex, complementary, holistic system with both "discrete life" and "sustained life" modes.

To further develop our distinctions, we now begin to use the term and capitalized word "Life" and will hereafter use Life to refer to this full meaning of "life as a unified whole." Given that choice, we will specify any other meaning by using terms like "organismal life," discrete life, the life of an individual, or when using "life" that is not capitalized as a proper noun.

One of the benefits of choosing Life as the basis of value is as a unifying foundation for ethics that has all humans, as interdependent with Life itself, equal and on the same team, in the same planetary boat. Expanding our area of recognition and concern for all Life shifts the focus onto cooperation and mutual benefits, and our shared relationship to Life and environmental life support come to the foreground. No organism is isolated and independent of its environmental Life support such that the entire habitat arena becomes compatible with the idea of "the commons." Environmental Life support systems such as the soils, oceans, or atmosphere are shared by all humans and all Life. And thus, for environmental commons like the atmosphere, and given the total necessity of a healthy, stable, and functional atmosphere, the emphasis naturally stays on maintaining and replenishing for mutual benefit rather than exploiting, consuming, or degrading.

Psychologically, and perhaps unfortunately, common cause can be promoted or inspired through an "us" versus "them" mentality. An individualistic framing would put the system boundary at "me" versus the environment, but as we've shown that has failed to protect the necessary Life support services that are not contained within this perspective. If, for some reason, we need to reinvent an idea of an "enemy," then from this unifying value of Life, we see that the enemy is not among us. The enemy is no one alive, as all living beings share Life and thus would "fight" on the same team with respect to protecting, valuing, promoting, and defending Life (again, with specific reference to life as a unified whole, not with reference to the discrete life of individuals, organisms—not even the life of nations, tribes, or families—and other artificially isolated subsets of Life). The enemy we humans all share is what we might call "systemic death"—*an end of all Life*, any threat to Life itself, and including extinction of humans as a species—as we discuss later, we see humans as having an essential role in the success and destiny of Life into the open future; thus, an end to humans is detrimental to Life.

We admit this may at first seem a big leap to convince people to consider equally any urge to preserve one's own life and the need to preserve Life itself. This may in fact be one of the hardest crux issues we address in this book—challenging the genetically and culturally deep-seated human sense of self and self-preservation, so as to extend these urges and instincts beyond the boundary of the individual and to include Life-support organically into our decision-making. We start this challenging effort here and continue for the rest of the book. A first step is to consider it as highly rational that anything one absolutely needs to live—the top priorities in Maslow's hierarchy of needs, physiological needs, such as oxygen, water, and food—these "things," and the processes and systems which provide them in sustained supply indefinitely, these are worthy of high value, care, and preservation.

As far as a shared enemy of all Life, this enemy of "systemic death" really exists and can be described, measured, observed, and characterized. It could be caused by the harshness of outer space, the ultimate outcome of entropy and the tendency to dissipation and disorganization if they are allowed to proceed to thermodynamic heat death, and potentially by impactors from outer space or catastrophe such as nuclear holocaust. Even a change as dramatic as arriving through anthropocentric climate change would hardly wipe out all Life, but rather make it quite difficult for continuing current human institutions and expectations.

We note, too, that our holistic conception of Life changes our views and discussions of death, systemic death, or perhaps Death with a capital D: an end of all Life. The concepts of death and Death are thus scalar or multi-scaled, similar to our multi-scale and complex ideas of life and Life. Individual death is very different than population death, species extinction, or planetary Death. This is consistent with our distinction between discrete life and sustained life. It also raises interesting questions: Is death of the individual really our enemy? This is perhaps an American or Western value more than a universal one. Many cultures view other features higher than individual life—honor, family, religion, etc.—and would readily trade an individual's life for those. In line with Life's broader and wider cycles, and the inevitability of human death, we do not suggest that the death of an individual is inherently bad ever and always. Many contextual factors come into play. It is really only capitalized, systemic, and ultimate Death—an end to all Life on Earth—that we suggest is inherently bad and something we humans could agree to fight against and oppose, as we increase our unity, solidarity, and cooperation for serving and sustaining Life.

One working analogy could be to consider three system states akin to phases of water—gas, liquid, and solid—where fluid liquid represents an adaptive and living state, and where the other two represent two forms of death. Life is balanced between too much rigidity (ice) and too much anarchy (gas). This implies there are two ways to die, to be out of balance in extreme in either direction.

A system in which entropic processes outpace self-organizing ones would find itself with too many connections, too little efficiency, and too much redundancy and dissipation, ending in the classic heat death early 20th century physicists feared. But, there is another imbalance that can occur when too few connections, too little redundancy, and too much efficiency lead to immobility and inflexibility. This results in a system or network that is frozen like a crystalline lattice and with no lifelike change, dynamism, or evolution, essentially displaying an "ice death." The shared goal of Life is to aim at the balanced, fluid, middle ground where Death is held at bay.

The second and last set of concepts and terms we define for the work of this chapter are "value" and "value system." The Stanford Encyclopedia of Philosophy describes "value system" under the heading of "social norms":

> In the theory of the socialized actor (Parsons, 1951), an individual action is equated with a choice among several alternatives. ... Order and stability are essentially socially derived phenomena, brought about by a common value system—the "cement" of society. The common values of a society are embodied in norms that, when conformed to, guarantee the orderly functioning and reproduction of the social system. In the Parsonian framework, norms are exogenous: how is a common value system created, and how it may change and why, are issues left unexplored. The most important question is rather how norms get to be followed, and what prompts rational egoists to abide by them. The theory of the socialized actor's answer is that people voluntarily adhere to the shared value system because it is introjected to form a constitutive element of the personality itself (Parsons, 1951).

From https://plato.stanford.edu/entries/social-norms/

This description of social norms and the closely related value system highlights the interdependence of the individual sense of self and the societal system of values and norms. We similarly see this dual role as important in the Great Change for sustainability, in that individuals and societies must transform in mutually reinforcing ways. The idea of value system as the "cement" of society is important too, although we might use a less industrial metaphor like "connective tissue."

The literature on value and value systems is large, and several value theory schools of thought exist, in which debates about intrinsic versus extrinsic value, and intrinsic versus instrumental value continue. Again from Stanford Encyclopedia of Philosophy:

> The term "value theory" is used in at least three different ways in philosophy. In its broadest sense, "value theory" is a catch-all label used to encompass all branches of moral philosophy, social and political philosophy, aesthetics, and sometimes feminist philosophy and the philosophy of religion — whatever areas of philosophy are deemed to encompass some "evaluative" aspect. In its

narrowest sense, "value theory" is used for a relatively narrow area of normative ethical theory particularly, but not exclusively, of concern to consequentialists. In this narrow sense, "value theory" is roughly synonymous with "axiology". Axiology can be thought of as primarily concerned with classifying what things are good, and how good they are. For instance, a traditional question of axiology concerns whether the objects of value are subjective psychological states, or objective states of the world.

But in a more useful sense, "value theory" designates the area of moral philosophy that is concerned with theoretical questions about value and goodness of all varieties—the theory of value. The theory of value, so construed, encompasses axiology, but also includes many other questions about the nature of value and its relation to other moral categories.

https://plato.stanford.edu/entries/value-theory/

The work to examine, classify, communicate about, and seek to build consensus around "what things are good, and how good they are," what the author describes above as axiology, is our main concern here. We build a case that Life is good, and that Life is among the "items" we humans should all classify as the ultimate good (other items may share this classification as ultimate good, but none can displace Life, otherwise Life is threatened and is in danger of nonexistence, at which point all other values and goods are meaningless).

We may also take a more biophysical approach, toward the question of social norms following the work of Odum (1971) who used the laws of thermodynamics to explain that social orders emerged as self-organized properties for better utilization of the society's available energies (with reward loops and penalties). He goes so far to as to propose a "Ten Commandments for the Energy Ethic for Survival of man in Nature" (p. 244). Yet another way to seek value aside from constructivist social sciences, and that is more consistent with our whole Life vision, stems from the field of environmental ethics, dating at least back to Leopold's work on the Land Ethic (see more on Leopold below).

The Oxford Dictionary defines value and values in different ways:

Value: "The regard that something is held to deserve; the importance, worth, or usefulness of something."

Values: "Principles or standards of behaviour; one's judgement of what is important in life."

https://en.oxforddictionaries.com/definition/value

These are the same meanings we will employ for *value*—seeking to show that Life deserves to be of the greatest importance and worth—and for *values*—seeking to show that Life deserves to inform our principles and standards of behavior and aid our judgment of what is most important.

Before describing the value system centered on Life, we describe how it fits into the overall strategy for solving the systemic sustainability crisis.

THE ANTICIPATORY STRATEGY OF SYSTEMIC CHANGE

A value system centered on Life is crucial to our book plan and strategy to solve the systemic human−environment crisis. In Fig. 2.1, we depict the change model, theory of change or working idea for how this book could fit into a larger process of scientific and societal change. This change model is analogous to other Great Changes in science and society where scientific advancements led to new ways to aid human life, and by extension, at least potentially, Life itself (Goerner et al., 2008). Some fairly recent (i.e., the 20th century) examples of major science-and-society advances include agricultural advances (nitrogen fertilizer), medicine (antibiotics), water and sanitation (water and waste water treatment including chlorination), and family planning [oral (and other) contraception]. In addition, certain science breakthroughs transformed science itself and people in science-and-society revolutions such as the Fossil Fuel Revolution, Industrial Revolution, Genetics Revolution, Information Revolution and others, which touched on all aspects of Life and how Life is organized. It is important to note that many if not all of these science major advances and revolutions raised the human carrying capacity on Earth (capacity of the environment to support human populations), at least temporarily if not permanently,

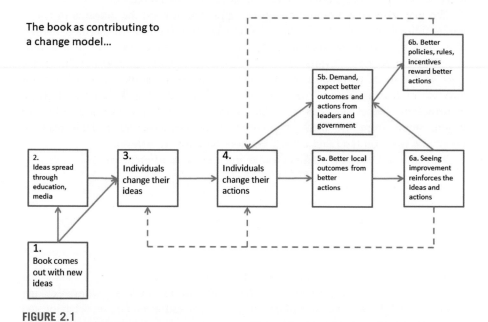

FIGURE 2.1

Our model of change, or theory of change, by which we imagine this book (and allied works and ideas of others) can contribute to societal change and eventual human−environment sustainability.

and increased complexity of the human—environment system. This increase in capacity and complexity then gave additional responsibilities associated with greater numbers of people, more complex science and society, and greater and more complex impacts on the environment.

As we said in Chapter 1, to solve our current suite of many environmental problems, we frame the sustainability and human—environment crisis as *systemic.* The many sub-crises (see Fig. 1.2 and Table 1.1) are then seen as "superficial" (relatively speaking) surface symptoms. We believe we cannot solve the individual subproblems in isolation—to attempt this is to validate the prior reductionist, specialist approach, which isolates and fragments problems and systems, and which we suggest is more aligned with the systemic cause of the problem than with a strategy for its solution. We also see evidence that this approach to focus on individual human—environment problems in isolation can result in a trap of solving one problem at the expense of creating one or more other problems—potentially worse problems—in some other human—environment area. This "robbing Peter to pay Paul" approach of unintended consequences, obviously, is not viable. Someone solved an isolated problem of a needed refrigerant by creating chlorofluorocarbons, which resulted in the far worse problem of destroying stratospheric ozone. Someone solved the shale gas problem by inventing fracking, only to threaten groundwater and aquifers, and release large amounts of methane, thus creating potentially deeper and longer lasting problems. Someone solved the problem of low-level infections inhibiting beef growth, production and profits, and innovated continual antibiotic use, thus creating "superbugs" resistant to antibiotics. Thus, we propose our holistic alternative is at the very least worth a thorough vetting and at best has a true potential to solve the systemic problem, and all associated symptoms of the ecological crisis, at once. This approach may not deter all unintended consequences, but not applying it guarantees that none will be anticipated.

One other critique of piece-meal approaches to solve isolated subset problems is that—especially as we know, or for those who do know or have a hunch, that our problem is systemic—failure to address the systemic crux problem is essentially procrastinating, "kicking the can down the road," waiting, assuming, and hoping someone else will do the hard work. Surely, we have the responsibility and it is worth the courage and conviction to face the deep root problem now.

Again, to recap from Chapter 1, from this systemic framing of the human—environment problem, we can set the goal, strategy, and approach to *solve the systemic root cause* of the human—environment problem and achieve lasting sustainability. And, in framing the solution case in similar holistic fashion, we see the solution as a change from a win—lose human—environment relation (in which environment degrades over time) to a win—win human—environment relation (in which environment improves over time).

To unify all people is one benefit of a revised foundation for science centered on the value of Life. Given that our human—environment problem is global, it impacts all people, and all people play parts in the collective human—environment relation. This connection to all humanity is relevant to all stages and all conceptual and applied aspects of our discussion. Our challenge is big, and we are thinking big—focusing our minds not only to reduce or to end environmental destruction, but also to solve the systemic human—environment problem rapidly and effectively.

We also propose a science revolution now based on the essential need for a specific future outcome: a successful transition to true sustainability. This follows Meadows' #1 and #2 sources of maximum leverage related to changing the paradigm of the system (see Table 1.4 in Chapter 1). Thus, we propose we should not wait for a slow Kuhnian science revolution like many of those in the past that can take one or more generations of scientists to slowly test and resolve major conceptual difficulties. The Kuhnian crisis in science is here and now (the current approaches are not working), and we need a more rapid paradigm shift to match the urgency of our planetary human—environment situation.

In order to achieve the new understanding of Life and its successful win—win relation with environment, we need to develop a new paradigm of Life and Life science. This is an anticipatory and novel approach to intentional use of "science itself" (all of science, not just a single study) to serve fundamental human needs and tackle a great challenge for humanity.

We proceed by these steps to the strategy (again, working from the "change model" in Fig. 2.1):

1. *Start with the end in mind*—Again, the #1 principle from Covey's Seven Habits of Highly Effective People. This anticipatory approach is essential, and it relates to the strategies of "completion backwards" and to be proactive. Another perspective is to think in terms of a "final cause" in the sense of the four Aristotelian causes (material, efficient, formal, and final causes). Aristotle brought an insight to the problem of causation that rested on four required aspects for any event to happen. First, the material itself that is acted upon must be present; second, an agent brings the material to action, but which is guided by the third element, a formal plan. Fourth, no event proceeds without its reason or justification for happening which is the final cause. Logicians often use an example of what "caused" a statue to appear in a Greek temple accordingly in which the block of marble is the material cause, the sculptor is the efficient cause, the blueprint/design in the sculptor's mind the formal cause, and the dedication of a new temple for Athena the final cause. Modern science puts most agency on the actor and to a lesser extent the material, but hardly ever on the purpose, which has been intentionally driven from and ignored by science. In Aristotle's vision, the statue would never have been built without the need for it. This perspective makes us question our motives and aspirations to ask not just "How?" but "Why?" A final compatible idea is to think in terms of what our ultimate goal in this work is, to set a course for what is most important and truly essential, even if this is very general and qualitative, what our gut wants, a hunch or intuition of the right direction in which to steer.

2. *Depict the goal and outcomes*—The next step is to describe the general ideal future, imagine it, add details, form a rich and clear vision of success. As above, this involves a future in which the human−environment systemic crisis has been solved once and for all, with a lasting and systemic win−win human−environment relation going forward. These aspects of the ideal endpoint and future are essential:
  a. All 11 of the crisis symptoms are solved.
  b. All such "vital signs" and Life support major metrics are self-regulating going forward, and no new systemic and Life-threatening problems are created.
  c. Human and environmental capital, capacity, and competencies are regenerated—healing of old wounds to environment (atmosphere, waters, soils, species, habitats, and more) and social cohesion (conflicts and animosity based on natural resource conflicts primarily, hopefully extending to conflicts of religion, ideology, and other divisions).
  d. All major vital signs and capacities are robust to disturbance and self-healing after disaster. Here, as in many areas, we can model success (Sustainer success) based on, and mimic for success, living systems. Thus, human systems will be like forests that recover after defoliation by insects, regrow after fire, and are resilient after other major disturbances.

In addition to the environmental vital signs (outcomes, indicators, top 11 symptoms), we also must ensure that social sustainability is achieved with social justice as integral at all phases. For this, we follow work of Julian Agyeman who defined "just sustainabilities." From his website julianagyeman.com, we read:

> Agyeman is originator of the concept of 'just sustainabilities,' the full integration of social justice and sustainability, defined as 'the need to ensure a better quality of life for all, now and into the future, in a just and equitable manner, whilst living within the limits of supporting ecosystems.'

At first read, this might appear impossible, idealistic, even utopian. However, all these features of ideal environmental success are what preindustrial living systems do naturally. The goal is not to romanticize nature blindly, as strife, hunger, bloodshed all existed in nature before humans and will continue (eating and being eaten is the way the ecosystem is organized). But, there is a "fitting in" that humans have lost and need to rediscover. In fact, there is no reason to believe that the intelligence of humans can't be a vital asset in designing, modeling, and constructing livable win−win socio-ecological relations, ones that would not be possible without the input of humans. The missing parts currently are seeing Life as model and mentor, and seeing that part of this process of setting a value basis centered on Life relates to efforts to do more biomimicry, and "ecomimicry," by which we learn to live in ways that enhance our environmental Life support systems. The tree in the forest makes the forest better. Or as Jacobs (2000) phrased it: "An ensemble grows rich on an environment that the ensemble itself made rich." (p. 60), which she observed in nature and reapplied to human-built environments such as cities.

Just as we do not romanticize nature blindly, we also do not romanticize or even accept as necessary our current state of affairs that includes systemic environmental degradation and thus degradation and devaluation of Life.

If at first read our proposals do appear impossible, idealistic, or utopian, then we ask you to consider reversing your model, as Elbow suggests. Our thesis in this book is that such a complete reversal is needed to provide the new perspective to allow for solutions and success. For example, instead of considering our descriptions of the ideal as impossible, try the opposite— consider that what we are experiencing now is impossible (to sustain), idealistic (to think it will all work out) and dystopian (if we become numb and accept it). In Chapter 5, we describe a founding scientific principle, radical empiricism, that fits with this radical shift in perspective. What if environmental degradation is the abnormal condition, and self-maintaining and self-improving environment is the norm? What if clean water, pure air, a stable atmosphere, abundant biodiversity, renewable energy, and other so-called "idealistic" or "utopian" dreams are common sense, readily achieved by basic natural Life systems, and every human's birth right?

To solve the crisis, we can mimic life which has solved this systemic issue—the net outcome of the relationship between life and environment over the long term is win—win. The last decades of research in systems ecology has identified a number of critical features that keep ecosystems functioning with high levels of complexity, organization, diversity, adaptability, and resilience. We can see this clearly because over the long term, the environment has improved in its capacity to support life.

3. *Work backward to realize the ideal outcomes—the "how" that leads to it.* Now that we have depicted the future goal, we can work backward to realize it. For this successful future vision to come true, we need science, technology, and culture that result in the win—win human—environment relationships and maintenance of our Life support systems. This is the "how" part—the realization and actualization, the building, operating, maintaining, doing, and the infrastructure for "business as usual" (i.e., a new form of business as usual that is successful, sustainable, and just). Given the emphasis on Life (again, life itself, life as a unified whole), life science must be elevated to a priority role among sciences. Thus, we need new ideas for biology, ecology, environmental science, and how these can all be integrated in holistic science and technology. Since the value of Life is shared by all humans, other sciences need to serve Life as well—chemistry in service to Life, physics in service to Life, engineering in service to Life, etc. It can be assumed that in a world where all humans value Life that the science and institutions of those humans will also value Life.

More specifically, the outcomes—whether direct and intended products or unintended byproducts and side-effects—must reverse damage, contribute to solving the systemic root cause of the crisis, or assist in maintaining high quality of human life and environmental Life support systems. The system is functionally locked together such that there are loopbacks that reinforce the overall goals.

4. *The "why" that motivates the "how."* For these transformed realizations of science, technology, culture, action, policies, and results to come true, we need an intentional and shared value of Life as the center and foundation of ethics and what is known to be good, right, important, essential, and prioritized. This is the paradigm of Meadows' leverage, which can then be codified and embodied in rules of the system and other systemic properties which must change (Meadow's goals, power to change system structure, structure of information flows, gain around positive feedback loops, etc. See Table 1.4). These are the societal norms and widely held assumptions of what is good, true, right, normal, accepted, and what informs or motivates business as usual. This is the "why" part; the final cause in Aristotelian parlance. Once it is set, the "how" parts of science, technology, and infrastructure are activated to achieve these goals and realize these values all day and every day.

SETTING LIFE AS THE BASIS OF VALUE

A journey of a thousand miles begins with a single step.
Chinese proverb from Tao Te Ching

Here, we take the first step on this long and complex "journey of a thousand miles"...setting Life as the basis of value for science and, by extension, for society. This is one form of the final answer to any long series of "Why?" questions, such as might be asked by an insistent and inquisitive child probing ever deeper to ask "Why?" again after each explanation of a proximate cause or intermediate step. We submit that a suitable answer will always be, To Serve Life. And we hope to show, and hope others will help to show, that this form of an ultimate answer (for which there may be multiple compatible additions), purpose, or value is compatible not only with the best of science, but also with the best of world religions and spiritual traditions. Such is the awesome unifying power of Life.

For this main section of this chapter, we follow Albert Schweitzer, Aldo Leopold, Robert Ulanowicz, and others that have already written beautifully and persuasively on these topics. We go further into the effort to restate, clarify, and corroborate these ethical tenets as key motivational impulse for the whole new system of ideas and new root metaphor. The ethical tenets proposed and used are ones that are corroborated or mutually reinforcing with scientific tenets, such as the scientific understanding of both life and Life, life—environment, and the needs of living systems for continued existence. To play the sequence forward again, based on Life value, we develop a science to serve Life. And, using that science, we develop a technology and other real-world applications to serve Life. From then on, the systemic cause of the 11 Life-threatening symptoms in Chapter 1 are replaced by values, science, technology, and culture of Life-enhancing solutions, infrastructure, business, and everyday activities.

This value system basis, in our analogy, is the rich and fertile soil needed to support, nurture, and feed the sprouting new holistic sciences, including the holistic life—environment science we describe in Chapters 3 through 6.

SCHWEITZER'S REVERENCE FOR LIFE

Albert Schweitzer was a medical doctor as well as a devout Christian. We see his approach to setting Life as the most fundamental value to be fully compatible to the approach we develop here. While his main approach—what he called "reverence for life"—came from the context of Christian theology and ethics, his knowledge of science and biology, and skills in philosophy and logic, contribute to the strength of his ideas which serve well as general and foundational values. We do not intend to imply that ideas of Life reverence are limited to Christian theology, but rather to underpin Schweitzer's background and motivation—in fact, we see him as a transcendental figure who could be at the core of any deep teachings.

In Chapter 1, we quoted Schweitzer (1965) and repeat a part of that quote here as encapsulating the main points:

> The essence of Goodness is: Preserve life, promote life, help life to achieve its highest destiny. The essence of Evil is: Destroy life, harm life, hamper the development of life.
> The fundamental principle of ethics, then, is reverence for life. (page 26)

To fit with this book, we would like to replace Schweitzer's term "life" with our term "Life" to signify Life as a unified whole, including Life as integrated with its essential environmental Life-support context and thus not to confuse the issue of the discrete life of an individual, organism, species, or any other fragmented subset of Life.

Later in the same book, "The Teaching of Reverence for Life," Schweitzer (1965) wrote:

> In the main, reverence for life dictates the same sort of behavior as the ethical principle of love. But reverence for life contains within itself the rationale of the commandment to love, and it calls for compassion for all creature life.

The riddle of extraordinary and exalted ethics that Schweitzer identifies is that true values should be viewed as ordinary and are to be carried out in everyday life—in other words, in the act of doing and becoming, there is no wasted effort, no cross purpose, no conflict of interest. This insight is not unique to Christian faith but is relevant to our present systemic crisis in that we confront the same essential issue—how to unify and harmonize the ultimate long-term ends of successful care and stewardship of Life with the proximate and short-term priorities, decisions, and actions of daily life.

Schweitzer's desire for an understanding of the deepest root of the matter motivated his sermon of 1919 (Schweitzer, 1969):

> We want to grasp the underlying principle of all ethics and use that principle as the supreme law from which all ethical actions can be derived. (p. 111)

This quest for "supreme law" shares much with scientific quests for universal laws of physics, chemistry, biology, ecology, and other fields—a shared consensus foundation and standard reference against which all new insights and questions can be tested for validity. In terms of a universal ethic, and linked to Schweitzer's mention of compassion and love, we are in the realm of the Golden Rule—another axiom which is both exalted and every day.

In his effort to "let reason speak" on ethics, Schweitzer notes that one of the best results of reason is, "...it can teach us a certain integrity and justice, and these things are more or less the recognized key to happiness." Merging reason's results of knowledge and happiness in wisdom, Schweitzer (1969) gets to the kernel of reverence for life:

> Desire for wisdom! Explore everything around you, penetrate to the furthest limits of human knowledge, and you will always come up against something inexplicable in the end. It is called life. It is a mystery so inexplicable that the knowledge of the educated and the ignorant is purely relative when contemplating it. (p. 114)

He goes on to say that the farmer contemplating his garden and trees and the scientist observing via microscope are in the same boat:

> Both are confronted with the riddle of life. One may be able to describe life in greater detail, but for both it remains equally inscrutable. All knowledge is, in the final analysis, knowledge of life. All realization is amazement at this riddle of life – a reverence for life in its infinite and yet ever-fresh manifestations. How amazing this coming into being, living and dying! How fantastic that in other existences something comes into being, passes away again, comes into being once more, and so forth from eternity to eternity! How can it be? We can do all things, and we can do nothing. For in all our wisdom we cannot create life. What we create is dead. (pp. 114–115)

He continues in poetic language that integrates the voices of heart and reason, spirituality, and science:

> If you study life deeply, looking with perceptive eyes, into the vast animated chaos of this creation, its profundity will seize you suddenly with dizziness. In

everything you recognize yourself. The tiny beetle that lies dead in your path — it was a living creature, struggling for existence like yourself, rejoicing in the sun like you, knowing fear and pain like you. And now it is no more than decaying matter — which is what you will be sooner or later, too. (p. 115)

It is difficult to paraphrase or to restate. . .his words are original and authentic:

I cannot but have reverence for all that is called life. I cannot avoid compassion for everything that is called life. That is the beginning and the foundation for morality. Once a man has experienced it and continues to do so — and he who has once experienced it will continue to do so — he is ethical. He carries his morality within him and can never lose it, for it continues to develop within him. He who has never experienced this has only a set of superficial principles. These theories have no root in him, they do not belong to him, and they fall off him. (pp. 115—116)

He goes on to say that reverence for life is essential to "building a new human race" and that "existence depends" on the transformation of individuals able to see this as a "true, inalienable ethic."

Written and spoken in a sermon very nearly 100 years ago, at the end of World War I, we hear these words loud and clear and just as relevant today. Schweitzer was a kind of Renaissance man, a whole person who seemed never to have been forced to choose between science and religion, a person fluent in languages of both heart and reason, a keen observer of both inner and outer worlds. Thus, we take it seriously, and find a solid corroboration for our efforts here, when Schweitzer notes that without experiencing "reverence for life" one "has only a set of superficial principles," and by extension no basis deep enough to provide a solid anchor and foundation for ethics and ethical actions. He also refers to "reverence for life" as "the foundation for morality" and states that without this foundation a person is morally weakened because other theories or ethics "have no root in him," "do not belong to him, and they fall off him." This fits with our proposal that Life has ultimate value, since if we threaten, weaken, or end Life, then nothing else of human concern can exist, much less matter; other "theories" or values are necessarily superficial in comparison and can only hold value when rooted in the value of Life.

ALDO LEOPOLD'S LAND ETHIC

Leopold (1949) put forth an ethic with the same emphasis on Life in the extensive sense of which we speak here. There is much written by Leopold, and about Leopold's "land ethic," and we mention just three main passages here.

First, to quote his succinct and eloquent ethic (Leopold, 1949):

A thing is right when it tends to preserve the integrity, stability, and beauty of the biotic community. It is wrong when it tends otherwise. (pp. 224—225)

In addition to aligning the right and good—the fundamental value—with Life, Leopold also refers to "the biotic community" thus clearly recognizing the essential interdependence of individuals and organisms as we do with "sustained life." He also explains how this ethic changes our view of ourselves:

> In short, a land ethic changes the role of *Homo sapiens* from conqueror of the land-community to plain member and citizen of it. It implies respect for his fellow-members, and also respect for the community as such. (p. 204)

Here, in addition to emphasis on biotic community, Leopold refers to "land-community," which, similar to his mention of "thinking like a mountain" below, shows his awareness and value of the integration of environment into the biotic community.

Next, we quote the story of Leopold's experience, perhaps like an epiphany, to which he attributes the inspiration and realization of his transformed view of life. Leopold (1949) tells of an experience in the mountains, after realizing what he and others saw at a distance was not a deer as they first assumed:

> When she climbed the bank toward us and shook out her tail, we realized our error: it was a wolf. A half-dozen others, evidently grown pups, sprang from the willows and all joined in a welcoming mêlée of wagging tails and playful maulings. What was literally a pile of wolves writhed and tumbled in the center of an open flat at the foot of our rimrock.
>
> In those days we had never heard of passing up a chance to kill a wolf. In a second we were pumping lead into the pack, but with more excitement than accuracy: how to aim a steep downhill shot is always confusing. When our rifles were empty, the old wolf was down, and a pup was dragging a leg into impassable slide-rocks.
>
> We reached the old wolf in time to watch a fierce green fire dying in her eyes. I realized then, and have known ever since, that there was something new to me in those eyes — something known only to her and to the mountain. I was young then and full of trigger-itch; I thought that because fewer wolves meant more deer, that no wolves would mean hunters' paradise. But after seeing the green fire die, I sensed that neither the wolf nor the mountain agreed with such a view. (p. 130)

He goes on to explain how—driven by the narrower view of life, and the short-term goal of eliminating competition—wolves were extirpated in most of the United States. He tells how the unchecked deer multiplied to levels unsustainable, beyond the carrying capacity of the land, and they died of population pressures and also damaged the vegetation and soils by overgrazing. The mountain itself suffered rapid erosion, due to the loss of a link in an interconnected system. The section with the quote above is titled "Thinking Like a Mountain," and Leopold's mention of the view and awareness of "the mountain" is akin to our search for an ethics and value system compatible with the perspective of sustained life and Life as a unified whole.

Our third and final reference to Leopold shows his holistic understanding of life cycles, and the integration of Life and environment. The book, *Round River* (Leopold, 1993), contained essays he was working on near the end of his life. Near the end of that book is a chapter titled "Round River" and subtitled "A Parable." It contains these quotes:

> One of the marvels of early Wisconsin was the Round River, a river that flowed into itself, and thus sped around and around in a never-ending circuit. Paul Bunyan discovered it, and the Bunyan saga tells how he floated many a log down its restless waters.

> No one has suspected Paul of speaking in parables, yet in this instance he did. Wisconsin not only *had* a round river, Wisconsin *is* one. The current is the stream of energy which flows out of the soil into plants, thence into animals, thence back into the soil in a never-ending circuit of life. 'Dust unto dust' is a desiccated version of the Round River concept. (p. 158)

He also wrote:

> A rock decays and forms soil. In the soil grows an oak, which bears an acorn, which feeds a squirrel, which feeds an Indian, who ultimately lays him down to his last sleep in the great tomb of man — to grow another oak...
> (pp. 159–160)

Leopold goes on to comment that the river is also analogous to a pipe line, and revised his diagram to show, that "the pipe line leaks at every joint," and "energy is side-tracked into branches" and finally notes that "...each animal and each plant is the 'intersection' of many pipe lines; the whole system is cross-connected" (Fig. 2.2). In this essay, we see how Leopold's ethics or value system was integrated with his intuitive and extensive understanding of energy flow networks that form the basis of modern-day ecosystem science. We likewise propose an ethic centered on the value of Life as supporting, and as supported by, a

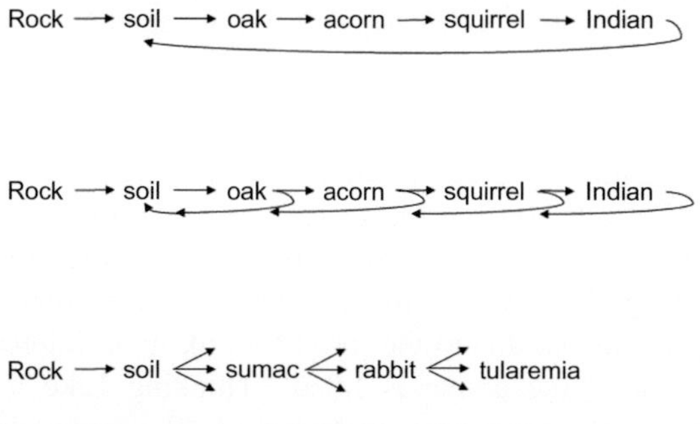

FIGURE 2.2

Diagrams redrawn from Leopold's Round River depicting both cycles and branches.

holistic ecological and Life science. In Chapters 3—6, we describe how ecological networks and other tools and insights help with the scientific paradigm shift needed.

In reference to Leopold's comments on the infancy of the science of ecology, and coinage of big words, it is interesting to note these history landmarks for ecological terms arising around the time of Leopold's writing:

1935—ecosystem—Tansley (1935)
1942—trophic-dynamic aspect of ecology—Lindeman (1942)
1950s—diversity and stability relationship—Elton (1958) and MacArthur (1955)
1969—strategy of ecosystem development—Odum (1969).

Also note that Tansley's (1935) first publication coining "ecosystem" echoed our unified life—environment view, shared by Leopold, and efforts for fundamentals and foundations, shared by Schweitzer:

> Though the organisms may claim our prime interest, when we are trying to think fundamentally, we cannot separate them from their special environments, with which they form one physical system. (p. 299)

Tansley's pathway to this integrated view of life—environment came via years of scientific research; Leopold's path included the experience with the dying wolf and efforts to understand and promote founding principles of conservation. Schweitzer (Kiernan, 1965) told a story of an early childhood experience when a friend dared him to kill an innocent bird with a slingshot, and he realized he could not due to his own internal strongest conviction; this he later articulated as "reverence for life." There are many pathways to this perspective and value system. Next, we hear the insights into fundamentals from another scientist.

ROBERT ULANOWICZ' ECOLOGICAL METAPHYSIC

Perhaps the contributions of Robert Ulanowicz do not speak to the primary value of Life as explicitly or directly as Schweitzer and Leopold, and we may have no quotes in which he says this verbatim, but he and his work very much show Life as a core tenet of his motivating values. In the preface to his book, *A Third Window*, Ulanowicz (2009b) shares that his goal has been to "describe an alternative approach to reality" in the form of "an ecological narrative" and an "ecological metaphysic." Much as we are proposing here, he has developed a coherent scientific narrative in his search "for new foundations upon which to build a rational description of nature," and he sees the need for a new science paradigm. If we can add even a single new supporting idea or concept as a new brick or beam to the edifice he has built in his ecological metaphysic, then it will be a humble contribution by comparison to his body of work. His work, in which we see the primary ambition for an alternative approach to reality to rejuvenate science to better serve humanity, pervades all aspects of our own work and thinking.

Ulanowicz shows his value of Life by three means that we address here. First, he shows a deep respect for nature and Life by his professional lifetime devotion of learning from living systems. While often reported in the language of science and mathematics, his body of published books, papers, and other works all reflect his passion for deep communion with Life in a sincere effort to understand. This respect and reverence is shown perhaps most clearly in his honest admissions and obvious openness to changing his own cherished ideas multiple times when lessons from his dialogue with nature proved it necessary. One example of this came when he realized that his major discovery of the ecosystem network principle of "ascendency" was not a unitary driving force able to explain the tendencies of ecosystem dynamics single-handedly. Submitting to the lessons of his encounters with complex living systems, he changed his views to put ascendency in context as one of two major tendencies of nature (the other being redundancy).

A second way that we read between the lines of Ulanowicz' work and infer a deep value of Life is in how he seeks to understand "agency"—what is responsible for causing events. This causal understanding is, of course, the gold standard of a valid scientific paradigm, and as we will see in coming chapters, we believe Ulanowicz and allied holistic scientists are pointing us in the right direction. But Ulanowicz has also done research and reported results in ways that show he takes responsibility for his own work, he is aware of his own agency, and he helps to make us all aware of our own avenues for empowerment as well. By demonstrating how "configurations of processes" at the focal level have agency—such as relationships in food webs (equally valid in human communities) rather than assigning all agency to the interaction of microscopic or subatomic particles as many reductionist and mechanistic approaches do—he has produced work and promoted ideas that inform and inspire our shared capacity to make change, to help others, and to do good in the world.

Lastly, Robert Ulanowicz has worked to apply lessons learned from study of living ecosystems to studies of human socioeconomic networks, thus promoting lessons of Life (in this case, complex ecosystems) from which we can learn and benefit if we abide by the same natural principles. In 2009, he and his coauthors wrote of their robustness index and information theoretic approach to understanding sustainability in networks (Ulanowicz et al., 2009a):

> The analysis provides heretofore missing theoretical justification for efforts to preserve biodiversity whenever systems have become too streamlined and efficient. Similar considerations should apply as well to economic systems, where fostering diversity among economic processes and currencies appears warranted in the face of over-development.

And later:

> It seems not unreasonable to assume that many of the same dynamics are at work in economics as structure ecosystems, and that, over "deep time," nature has solved many of the developmental problems for ecosystems that still beset human economies.

This effort to respect Life as model and mentor is another form of leading by example in which we see, follow and wish to appreciate Ulanowicz' value of Life as well as his own creativity, insights, and many scientific and philosophical contributions.

As we said above, and in both *A Third Window* and a paper on the topic (Ulanowicz, 1999—Life after Newton: An ecological metaphysic), one succinct phrase for Ulanowicz' main goal is "an ecological metaphysic." The Oxford Dictionary of Philosophy (Blackburn, 2016) describes metaphysics as "…any enquiry that raises questions about reality that lie beyond or behind those capable of being tackled by the methods of science." It goes on to say that "metaphysics…tends to become concerned more with the presuppositions of scientific thought, or of thought in general…"

Compatible with what Ulanowicz' works show for an ecological metaphysic, we hope that this chapter has made a solid case that Life can serve in this arena of the presuppositions of science and thought. As we laid out above in the theory of change, or evolutionary scenario, we imagine for how we can achieve true environmental sustainability, and by admitting that other valid worldviews exist (in Chapter 1), we do not suggest "Life as the basis of value" as an absolute truth or universal metaphysic. We do, however, assert that it must be considered as necessary if we wish to survive, thrive, and sustain our existence as a real life form and biophysical species living in intimate interdependence with the Earth as Life support system.

Toward summary and synthesis of this section, we note that the works of Schweitzer, Leopold, and Ulanowicz all refer to *cycles* of life, and we see this common idea to be of great importance for the foundations of value and science we propose to aid sustainability. While the Circle of Life is a common theme even echoed in popular culture and Disney movies, we feel it is hugely underappreciated. We focus on it now as a central aspect of a value system necessary to Life and human culture, and we will see the concept of cycles again as related to scientific formalisms (Chapters 3−6) and as key to policies, actions, and sustained system designs (Chapter 8). To recap two key references above, and to add one for Ulanowicz:

1. Schweitzer (1969) wrote: "How fantastic that in other existences something comes into being, passes away again, comes into being once more, and so forth from eternity to eternity!"
2. Leopold (1993) wrote: "In the soil grows an oak, which bears an acorn, which feeds a squirrel, which feeds an Indian, who ultimately lays him down to his last sleep in the great tomb of man − to grow another oak…"
3. Ulanowicz (2009b) wrote of autocatalytic loops, closed cycles of energy, matter, or influence in natural food webs and ecosystems: "The action of autocatalytic feedback tends to *import* the environment into the system or, alternatively, *embeds* the system into its environment."

The concept of the essential cyclic nature of Life (and death) is profound—this serves to unify Life (and death) and environment into a single entity, and from this unification makes fully equal the value of living beings and their fundamental Life support context. As long as the circle is unbroken, Life remains unified with environment. This universal life−environment cycle also brings a unifying quality to all humans across all races, nations, and ethnicities, and between all life forms. Referring back to Fig. 1.1, this helps prevent the fragmentation of Life from environment and subsequent devaluation of the environment relative to Life that we see as the first step that yields the systemic crisis and associated "tragedy of the commons" (Hardin, 1968) we now confront.

If it feels we have gone on too long, then please note we have not covered many other workers and their corroborating views on the value of Life. No study of this topic would be complete without inclusion of the work of Naess (1989) on "deep ecology" and the "ecological self" and Wendell Berry's writing on health as a property of a community. Christopher Uhl has written on "developing ecological consciousness," a closely allied concept and one of equal importance. The Dalai Lama has stated that "What we need…is compassion at every level— including for the planet" (Goleman, 2015), thus adding Buddhist perspectives of awareness, mindfulness, and compassion to the modes by which we can and must value Life and environment. Roderick Nash (1989) has depicted the history and evolution of environmental ethics in his book, *The Rights of Nature*. We cannot cover these and many other compatible thinkers in depth here, but we provide readings in the references. We hope we have whetted the reader's appetite for further exploration of Life as linked to fundamental value.

As a step toward transition to Chapter 3, in which we again borrow from Ulanowicz to describe a new holistic science, a few last thoughts on his ecological metaphysic. We note that we could have included in this chapter on Life as the basis of value important complementary work by Bernard Patten, as he, too, has contributed to ecosystem and network science which supports a primary ethical valuation of Life as a unified whole. Ulanowicz' and Patten's work in ecological network analysis, systems modeling, sustainability, and related fields shows repeatedly that ecosystems—as represented by food webs, stocks and flows of energy, carbon, and nitrogen (and at times other biophysical "currencies")—are bona fide, holistic life entities. Their work demonstrates that essential Life properties exhibited by ecosystem networks (i.e., autocatalysis, indirect mutualism, ascendency, network synergism) are not observable at reduced scales (i.e., the organism, species, predator−prey interacting pairs, or other isolated subsets of Life). They thus teach that ecosystems must be treated as fundamental units of study, and units of Life, and cannot be treated as epiphenomena explainable by dynamics of interacting subset parts. But, we reserve further discussion of Patten's contributions for Chapters 3−6 on holistic science.

The ecological metaphysic Ulanowicz sees as general and robust enough to be valid for many scientific fields even beyond ecology is based on three axioms (Ulanowicz, 2009b): (1) systems are vulnerable to disruption by chance events, (2) a process, mediated by other processes, is capable of influencing itself, and (3) systems have different histories, and some aspects of unique histories are recorded in material configurations. Note how these axioms are more lifelike than mechanism-like—as being applied to all of reality, they correlate more with a root metaphor in which the world's systems have inherent lifelike capacities for responding to disruptions and events, for self-reference and self-action, and for dynamic change over time, as opposed to a root metaphor of a mechanistic reality of dead interacting matter/energy components. Emphasizing the ubiquitous networks of relationships in ecosystems, Ulanowicz (2009b) shows how mutualism is ontologically prior to competition, which supports a view of an inherently mutualistic aspect to the original and fundamental nature of Life. Ulanowicz' canon of work provides results, examples, evidence, case studies, and analogies that corroborate the view of community-ecosystem networks as bona fide, holistic, irreducible units of life. Via sustained study of this level of organization, his work and that of allied systems ecologists verify that wholes are "greater than the sum of their parts."

Putting these together, Ulanowicz promotes ecology and life science—his "third window" perspective—as equal to, and perhaps now more important than, either of the prior windows/perspectives of Newtonian mechanics and Darwinian evolution. We move next to examine several key vistas from this perspective that we view as crucial to our future.

Holistic science of life—environment — mutualistic interfaces

3

The goal of the next section of the book, which includes this chapter and the next two chapters, is to describe a new form of holistic science, a science grounded in the value basis of Chapter 2, in which we have set Life in all its forms—Life itself, Life as a unified whole, and Life as integrated with its essential life support environment—as the primary basis of human value; we also seek a science able to solve the systemic human—environment crisis at its root. We address this holistic science from three perspectives related to concepts and principles.

In this chapter, we present an introduction and overview—our starting point, working assumptions, and modeling strategies for Life—environment as a unified whole and how and why we have gotten to this perspective. In Chapter 4, we address major context issues to help frame studies of Life—the origin of Life and the question "What is Life?" This includes a chronology and story—starting even before the origin of life; we share our hypothetical and relational narrative of how Life originated. This story starts with a seamless wholeness that gave rise to Life and the Life—environment relation via a series of developments and co-developments through bifurcations that continue to unify life-and-environment in mutually beneficial synergy even as aspects of each realm differentiate, diversify, and increase in complexity.

In Chapter 5, we propose the six core principles of the science we seek and are developing, including a science grounded in Life as value basis, that is anticipatory and oriented toward a future successful transition to sustainable human—environment relations, that balances and synergizes holism and reductionism, is radically empirical, and other founding principles.

As previously described, we propose a new science oriented to serve Life. We have also made the case that the primarily reductionistic and mechanistic science of the present and past, as excellent as it has been for many advancements and powerful human capacities, must be held responsible at least in part for the current global ecological crisis. Thus, our existing science has not accomplished what we see now as job #1 (following Leopold, Schweitzer, and others)—to promote, sustain, and enhance Life. Once we get the basic Life necessities covered, and once environmental life support is inherently being protected, nurtured, and enhanced, as required when Life is the ultimate value, then we can move on to other human values and human projects.

Foundations for Sustainability. DOI: https://doi.org/10.1016/B978-0-12-811460-5.00003-0

INTRODUCTION AND OVERVIEW

As we attempt to do always, we start by acknowledging prior holistic writers who have helped prepare the ground for our current project. A Billings' quote (1952) helps with the context of our approach:

> Since the environment is a complex it has been customary for ecologists and plant physiologists to break it up arbitrarily into factors and to study the effect of such single factors on the plant. This is a somewhat artificial, but probably necessary, method of attack, since in nature it is almost impossible for one factor to change without affecting others. Yet if such an analytical approach is followed, each factor must be evaluated in relation to all of the other factors, and the analysis must be followed by a synthesis of the total results. (p. 252)

The approach in the first part of Billings' quote is familiar: to break "it" up—where "it" can be the environment, some aspect of plants or Life, or really any scientific system of study—arbitrarily into smaller separate and single factors. This is the widely shared method of "reductionist science," which is the mainstream and primarily analytical approach used in almost all branches of science far beyond just ecology and plant physiology. We have lost sight of the fact that science and analysis have become essentially the same idea. The Merriam Webster Dictionary gives the first definition of analysis as "A detailed examination of anything complex in order to understand its nature or to determine its essential features: a thorough study doing a careful analysis of the problem" and the second definition as "Separation of a whole into its component parts." This same source describes the origin of the word analysis as coming from the New Latin, and from the Greek, "from analyein to break up, from ana + lyein to loosen."

We submit that *analysis* cannot constitute a full scientific process, not even if the breaking apart into a reduced number of factors is done strategically and intentionally rather than arbitrarily. The second half of the full process Billings describes—"the synthesis of the total results"—is sorely neglected and rarely done. Given that this bias toward analysis and breaking up of environment and Life factors has gone on for decades and even centuries, there is much overdue synthesis needed to stay true to this fuller vision of Life and environmental science as indicated by Billings. We hope this book and works of holistic colleagues can stimulate a longer process—perhaps, a matching synthesis period of similar length stretching for decades or centuries.

After years of studying the multiple and entangled symptoms of the human—environment crisis, we have developed the framing that fragmentation (stemming in part from science biased heavily toward analysis) is at the root cause of the systemic crisis we now experience, and thus, we see holism at the root of the solution. Fig. 1.1 in Chapter 1, depicts this with "System of Ideas" at the center of the bullseye diagram, the locus of understanding the root cause, and the caption describes the scientific paradigm as an integral part of this problematic shared cultural system of ideas: "In this science paradigm, life is separated from environment thus severing the unity of life and life support systems conceptually and scientifically." In Chapter 1, we described how, from this central problem, bias, or imbalance, values associated with this fragmentation, and especially a lower value of environment relative to life, then influence actions, which we assert then lead to the many superficial (but still profound) symptoms of life support collapse that we see everywhere.

The fragmentation emerges from the very way we think about and analyze the world as stated above. By treating a system as only a set of parts, we miss critical relations. It sets up an "us" versus "them" mentality in which that which is not "us" surrounds us but is separate from us. Again, quoting from Wendell Berry:

> A minor problem, perhaps, is the tendency of materialism to objectify the world, dividing it from the "objective observer" who studies it. The world thus becomes "the environment," ... which means "surroundings," a place that one is in but not of. The question raised by this objectifying procedure and its vocabulary is whether the problems of conservation can be accurately defined by an objective observer who observes at an intellectual remove, forgetting that he eats, drinks, and breathes the so-called environment.
>
> **Berry (2001, pp. 25–26)**

Another way of looking at this is to examine the potential for downside or negative side effects associated with a key aspect of the strength of modern reductionistic science—the narrow, extreme, and laser-like focus on understanding specific relationships, processes, and phenomena in nature. While this approach has led to an explosion in knowledge gained about how various individual aspects of natural systems work and behave, when this narrow focus on subsets of nature treated in isolation is translated into technology development and problem solving, a new issue arises. It becomes possible and even commonplace that a well-meaning scientist or engineer (or a team, or an entire disciplinary field) can solve one problem with seeming brilliant success, but only at the expense of creating one or more worse problems somewhere else. This is the proverbial pushing down a bump in a rug only to see it reappear elsewhere in another location. Removing sulfur from coal before combustion ameliorates an air pollution problem but creates a solid waste or water problem. Only a transformative technology, such as alternative energies, could fully satisfy the situation. Because the initial science involved a reduction of the system of study that necessarily cuts out most of the world in order to focus, any liabilities, damage, or unintended consequences that may arise have also been truncated from view and are thus invisible and deemed unimportant.

When "successful" solutions created in isolation are also heavily rewarded by financial gain and professional stature (publications, tenure, promotion, awards, etc.), we get a positive feedback driving the system growth toward hyper-specialization. This then leads to increasingly fine-grained reductionism, ever narrower fields of focus, and ever greater chance that the big picture is ignored and left on its negative trend toward being the forgotten commons, the wasteland of unintended consequences, the dumping ground of the "messy" aspects of problems, which are easier to be ignored than explained, all associated with the scientifically and socially removed realm of the "externalities" in economics. By repeating the process of isolating microscale science studies to those variables that can be controlled, and by then building engineering and technological solutions leading to greater micro-control, we inadvertently leave a huge mess for someone else to figure out—scientifically, socially, environmentally, and economically.

We now focus on a new science paradigm that mends the key mental fragmentation of life from environment, contributes to a better system of ideas grounded in Life as primary value, influences better actions, and with the hope, hypothesis, prediction, and anticipation of better impacts and outcomes toward healing current damage and fostering long-term sustainability. We see this proposed holistic Life-centered science as uniquely able to help unify disciplines and fields that have been splintered by hyper-specialization and to provide pragmatic tools for synthesis as balance and antidote to unfettered analysis.

The fragmentation of life from environment sets up a zero-sum scenario that requires, or creates the appearance of, winners and losers. If there are fixed resources, then our taking leaves less for others. This dilemma has been acknowledged for over 150 years. Perhaps, the first scientific treatise on this was George Perkins Marsh's *Man and Nature* who wrote in 1864, "A certain measure of transformation of terrestrial surface, of suppression of natural, and stimulation of artificially modified productivity becomes necessary. This measure man has unfortunately exceeded." Recognition that we have taken and modified the Earth too much is the first step to change the pattern, but rather than simply taking less, an integrated solution would refute the premise that the Earth must lose for us to win. Similarly, in Chapter 1, we framed the current problem this way, which bears repeating:

> The fundamental, net human—environment relationship is antagonistic or win-lose.

The crux of the problem and solution is a shift between two scientific views of the fundamental relationship between life and environment.

However, even with fixed resources, our using them can enhance their function for others. Thus, we propose the holistic summary of our global ecological solution, and the desired outcome we seek to assist is a future in which:

> The fundamental, net human—environment relationship is mutualistic or win-win.

Following achievement of this systemic solution, human actions would naturally and consistently result in *improvement* in the environment and Life support systems over time, reversing the trend of widespread, chronic, and systemic environmental degradation we see now. Healing the fragmentation removes the inherent confrontation that otherwise dominates perception translated into action. This qualitative change from win—lose to win—win human relation to environment may sound simple, but the full implications are profound; and so we next dig deeper into this issue.

BACK TO THE DRAWING BOARD—SEEING IS KEY TO ACHIEVING

As we begin to understand the role and responsibility of science as both partial cause and potential solution for our current Life-threatening crisis, we must examine many of the hidden or tacit assumptions in our scientific enterprise. This examination includes study of founding ideas codified into our primary textbooks and educational curricula all the way up to well-funded research labs. It also addresses everyday working assumptions, behaviors, activities, and choices by people inside and outside of science.

We can also observe our situation and circumstances—the world we have in part created through our science and technology—with newly focused holistic eyes. One way to summarize our crisis now is that humans threaten the environmental life support capacity for our own species and many other species. At the very least, one could expect motivation for repairing environmental damage because of this profound dependency we humans have for all resources often referred to nowadays as ecosystem services. This partition and lack of recognition of the prominence of ecosystems at the base of the pyramid is even evident in the current thinking around sustainable development. A common metaphor is that of a three-legged stool or overlapping circles in a Venn diagram (see Fig. 3.1A), where the three domains are social, economic, and ecological. The image tries to convey the idea that the only sustainable solutions are the ones at the center where all circles overlap. However, such a perception of three circles reinforces a concept that there are areas of each domain that do not overlap, that somehow there can be economic activity completed divorced from social or ecological aspects. Of course, this is crazy and not possible. Therefore, a more appropriate metaphor is one of nested circles with ecology at the base, human society a subset of that, and the economic activity a subset within that (see Fig. 3.1B). There are social aspects that are noneconomic such as family and community relations, volunteering, worshipping, recreating, meditating, etc. However, there are no economic activities that are not part of society. The same goes for the primacy of the ecological domain.

Upon serious reflection, this negative circumstance (global crisis, damage to environmental Life support) is highly problematic and paradoxical *scientifically*. Mainstream biological and economic theories hold that self-interested and competitive behavior of individuals—the prevailing cultural principle which has radiated outward from the prevailing scientific paradigm—naturally leads to betterment for individuals, species, and for living communities as a whole. Within

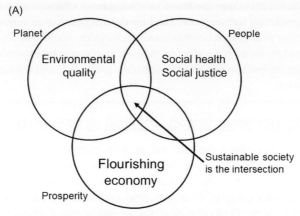

(A)

Planet

People

Environmental
quality

Social health
Social justice

Sustainable society
is the intersection

Flourishing
economy

Prosperity

**Overlapping aspects of a
sustainable society**

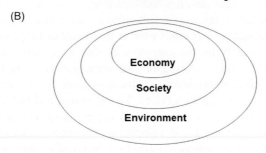

(B)

Economy

Society

Environment

A nested view

FIGURE 3.1

(A) Overlapping realms of sustainability. (B) Nested realms of a sustainable society.

evolution, the explanatory theory of natural selection holds that those organisms that are better adapted to the environment survive and reproduce more than variants of the same species that are less fit. While a core aspect of this theory addresses the relationship of an organism or species to its environment in general, and this natural selection can involve variation in fitness to abiotic variables such as temperature, moisture, physical disturbance, pH, and many other factors, it is often the case that the most dynamic and most challenging aspects of the environment are living actors. Thus, an integral and primary focus is on competitive relationships—innovations are selected that lead to competitive advantage of one life form over another, such improvement spreads by still more competition, and all life forms become better adapted to the environment as a result. Within social and economic realms, competition is similarly viewed as the fundamental relationship that leads individuals to strive to better themselves in the contest for social and economic rewards and again is assumed to lead naturally to improvement in business, technology, and efficiency for all.

Another perspective on this is to note that the working assumption is that betterment of individuals scales up to higher levels of organization, such as to populations, communities, ecosystems, and all of Life. This is akin to saying that the life units at these varying scales—from the individual up to the biosphere—all behave the same or in fundamentally similar ways. Another possible take on these linked assumptions is that many seem to believe that the smaller scale units are the only ones that really matter for explanation and that various forms of individuals, acting like "elementary particles," are the only unit or scale of analysis that has value or meaning. It ignores heterogeneity and hierarchy. This is another core tenet of reductionism—that phenomena are best explained by behaviors of the subsets and smaller units at levels below. Ulanowicz (1997, 2009a,b) has critiqued this working assumption of explanation coming from levels below and found it to have deep faults in his revolutionary offering of a philosophy of science grounded in ecology.

The principle of improvement through competition is thus treated as universal in the current dominant worldview and as equally applicable to nonhuman species and natural selection as well as humans in social and cultural interactions. Familiar social/economic examples include the way competitive grants are made for scientific research and other project funding, and the competitive bids process for construction and development work.

But the increasing and irrefutable evidence of systemic environmental crisis presents a serious paradox and contradiction for both realms in which this theory has been assumed to hold, for any claim that it is universally and generally true, and for any claim that it can be uniformly applied across all levels and scales of organization. Our current circumstances indicate clearly that some major explanatory principles must be missing that can account for the negative side effects in the case where humans have "won" competition against other species, and in which industrial nations and economies appear to be "winning" in competition against other socioeconomic systems.

We previously described this paradox this way (Fiscus et al., 2012):

> ... the multiplicity of chronic and systemic symptoms—arising from industrial societies and economies that predominantly emphasize competitive action for self-interest—suggests a deep problem. From [the] holistic perspective, the most basic relationship of humans to environment appears clearly dysfunctional and suicidal. The net relationship in the short-term is that humans win (gain resources, grow, develop, etc.) at the expense of the environment (environment deteriorates and loses, resources are depleted, wastes accumulate). However, since we are dependent on the environment for life (and in fact not even clearly separable; see below), if this relation continues in the long-term, then humans eventually lose, too. (p. 47)

One also can view this as a mismatch between proximate goals or values (e.g., to gain resources, to spur economic development in the short term) and ultimate goals or values (e.g., to sustain human life and to provide essential life support services for all people over the long term). Currently, society often has a hard time distinguishing between means and ends. For example, a rational interpretation of economic growth is that it is a means to an end but has instead become an end in itself. This could be partly due to ignoring Aristotle's Final Cause, which is the reason that something occurs, without which there is no point in the action. Also, it is just more convenient and lazy to turn ends into means since it avoids deeper and harder questions about the objectives being pursued. Raskin (2017), who used multiple scenarios of the future as a way to understand our choices (and their implications) now, wrote of one positive future he and colleagues called the New Paradigm within a path called the Great Transition. In the positive future after the Great Transition, Raskin wrote, "economies would be understood as the proximate means to the ultimate ends of vibrant lives, harmonious societies, qualitative, not quantitative, development" (p. 49). Wise distinction combined with clear realization of proximate and ultimate values is a generally useful approach as we have seen in Chapter 2 and again here. It appears to be an essential feature of *Homo sapiens* as we mature into a more beneficial relationship with our environmental home.

The current situation matches key aspects of the conditions prior to a scientific paradigm shift as described by Kuhn (1962) in his theory of the structure of scientific revolutions. First, there are anomalies, followed by persistent anomalies that lead to a full-blown crisis in one or more fields of science. Second, there must be available an alternative theory or paradigm that resolves anomalies and crises and that fits better in explanation of nature. Kuhn (1962) wrote:

> ... in all these cases [multiple past major scientific revolutions] except that of Newton the awareness of anomaly had lasted so long and penetrated so deep that one can appropriately describe the fields affected by it as in a state of growing crisis. Because it demands large-scale paradigm destruction and major shifts in the problems and techniques of normal science, the emergence of new theories is generally preceded by a period of pronounced professional insecurity. As one might expect, that insecurity is generated by the persistent failure of the puzzles of normal science to come out as they should. Failure of the existing rules is the prelude to a search for new ones. (p. 68)

And later:

> ... once it has achieved the status of paradigm, a scientific theory is declared invalid only if an alternate candidate is available to take its place ....

> ... the act of judgment that leads scientists to reject a previously accepted theory is always based on more than a comparison of that theory with the world. The decision to reject one paradigm is always simultaneously the decision to accept another, and the judgment leading to that decision involves the comparison of both paradigms with nature *and* with each other. (p. 77)

We see the current mounting crisis with our environmental Life support systems, and the failure of "normal science," mainstream mechanistic and reductionist science to solve this crisis, as the necessary preconditions for the next major scientific revolution. Kuhn's (1962) work provides additional useful framing where he described "normal science." He wrote of normal science research as:

> ... a strenuous and devoted attempt to force nature into the conceptual boxes supplied by professional education .... Normal science, for example, often suppresses fundamental novelties because they are necessarily subversive of its basic commitments .... Sometimes a normal problem, one that ought to be solvable by known rules and procedures, resists the reiterated onslaught of the ablest members of the group within whose competence it falls. (p. 5)

And

> ... when ... the profession can no longer evade anomalies that subvert the existing tradition of scientific practice—then begin the extraordinary investigations that lead the profession at last to a new set of commitments, a new basis for the practice of science. (p. 6)

Ulanowicz (1997, 1999a,b, 2009a,b) is the leading developer and proponent of the alternative paradigm that we work to help develop here. We seek to help solve problems in the real world as a means to develop and test aspects of the new theory and paradigm. For this problem-solving effort, we turn to the understanding generated by centuries of research and thought on ecology and nature.

A crucial step is to observe that Life has achieved what we seek to solve ultimately, our current real-world crisis. The Life−environment relation is inherently win−win on balance, such that the environment has improved over time as influenced by the operation of Life systems. We have already mentioned briefly that key forms of evidence for this win−win relationship include soils, the oxygen atmosphere, and the ozone layer. We next provide more in-depth description of each of these systemic improvements in Life support and then discuss how to understand this powerful and ideal Life−environment relation.

While we see the Life−environment relation as hugely important—almost miraculous—it seems to have been profoundly underappreciated that, over the long term, the environment improves in terms of its quality of Life support and carrying capacity for Life. The evidence of this directional change is in the increased diversity, energy density, and structural and functional complexity (at chemical molecular, organismal, and system organization levels) to name a few.

Other scientific narratives within the dominant mainstream focus on how unexpected it is that life emerged from the "primordial ooze," bootstrapped itself into existence, managed self-organization and increasing order and complexity despite the second law, and survived major cataclysms like impactors and climate change over geologic time. While this is an interesting story with much truth and power, we see another wonder that gets much less attention and examination—how Life has managed over billions of years not only to avoid fouling its planetary nest but also to feather that nest so beautifully. Our theory fits more with

Kauffman's (1995) ideas in his book, *At Home in the Universe*: life is not random and unexpected, but very much to be expected when the self-organizing tendency of the universe is fully understood and appreciated. By extension, we propose that a win—win Life—environment relationship is likewise to be expected.

As we address the specifics next, such as the oxygen atmosphere, we admit that, from the perspective of some early species and groups, the presence of oxygen originally was a great "fouling" and much life died off as this more reactive element increased and proliferated. In this sense, there were distinctive groups of winners and losers. However, when we intentionally adopt the orientation to focus on Life as a unified whole, we assert that the invention of the oxygen atmosphere was of greater overall benefit than harm. Furthermore, recall that we use Leopold and Schweitzer's stance of value relative to all Life, Life itself, Life as a unified whole. Schweitzer (1965) wrote, "The essence of Goodness is: Preserve life, promote life, help life to achieve its highest destiny." And Leopold (1949) wrote, "A thing is right when it tends to preserve the integrity, stability, and beauty of the biotic community." One additional detail here is that many of the anaerobic microbes for which oxygen is toxic still exist and play important Life roles even today—they have found many suitable environments with little or no oxygen.

Several major sets of evidence serve to illustrate the point of Life—environment mutualism, but one must be open to reflect deeply on factors which may have become common to the point of being taken for granted as merely "the way things are."

THE OXYGEN ATMOSPHERE

First, the oxygenated atmosphere of Earth is unique compared to other planets (Lovelock, 1988). The unique atmosphere, as it developed over time, has served to make higher rates of energy use and metabolism possible (Swenson, 1989; Goerner, 1999). This far-from-equilibrium atmosphere also makes it possible that life forms like mammals can arise and exist, as the energy required for mammalian respiration and metabolism depends on the high oxygen content and high redox potential of the environment. To say this another way—if Life's actions did not result in a net positive impact in the environment, then we would have no oxygen atmosphere, no mammals, and no humans.

The composition of Earth's atmosphere in early stages of formation was largely carbon dioxide. The atmosphere is believed to have formed from intense volcanic activity which ejected carbon dioxide as well as water vapor, ammonia, and methane. Life arose on Earth under these conditions (see below for our ecosystemic and holistic hypothetical scenario of the origin of Life and the Life—environment relation). Without oxygen, the heterotrophic life energetics utilized anaerobic metabolism such as glycolysis. This has very low and limited efficiency in generating ATP from each molecule. Over time, perhaps around 2.45 billion years ago, additional energy capture pathways evolved which led to photosynthesis (probably following on from chemosynthesis as a precursor); the direct transformation of solar photons into stored chemical energy bonds. As a result of this process, carbon was actively removed from the atmosphere and oxygen was emitted as a by-product. Again, over long time scales, the chemical composition of the atmosphere changed noticeably. This is referred to as the Great Oxygenation Event. Oxygen is a highly reactive molecule and its presence opened the doors for additional biochemical advances, notably oxidative or aerobic respiration. This new pathway was much more efficient at generating ATP from each molecule allowing the organism much greater resources for growth and development. As described by Hall (2017):

> If one takes a broad view of evolution any time a new technology with a high energy return happens along in the evolution of life there will be an explosion of life forms using this technology. For example, ... Life was abundant and diverse, but none of it used oxygen (as we do) as a terminal electron acceptor because there was no oxygen available. Life operated on fermentation, generating energy-rich alcohol as an unusable by-product. But once land plants evolved and generated free oxygen as a by-product of photosynthesis, other organisms were able to utilize this oxygen and increase their own utilization of their own food (such as plant sugars) by about a factor of 4.

Organismal development abounded with this new-found energy, leading to thermoregulation, increased mobility, and neurological and biochemical complexity to name a few. And, this development eventually led to mammals and to humans.

Again, examining both the positive and negative aspects, we could say that it was a form of "creative destruction" in that the old Life systems gave way and opened up for new possibilities.

The story of the oxygen atmosphere is important for several reasons. First, it is an example of how complex systems evolve and develop to more complex ones along particular path-dependent trajectories. Often, these are not based on the immediate survival aspects that would be selected for. The oxygen atmosphere, in fact, was quite harmful to a great number of species leading to one of the major extinction events. Yet, the unintended by-product of oxygen and the new development pathway that was found was an exaptation in the language of Stuart Kauffman (2000). It was an aspect of evolution that was not immediately relevant yet became adopted and even critical in other conditions.

Second, the oxygen atmosphere demonstrates how Life changes the environmental conditions that further enable Life to thrive and flourish. This co-development and self-regulatory feature of Earth as a complex system was noted by James Lovelock leading to his Gaia Hypothesis. Without going into the details of the entire theory, it is useful to point out that the origins for this came from the question regarding atmospheric compositions. The story goes that Prof. Lovelock was approached in the 1960s by space scientists—before there was a field of astrobiology—with the question about whether life exists on other planets. During this early period of space exploration, the space scientists' preferred answer would be uncertainty that would lead to human missions to other destinations, which can capture the public's imagination, funding dollars, and fame. However, after some consideration, Prof. Lovelock provided his insight that an answer can be found simply by looking at the atmosphere of the celestial body in question. His theory was that life would leave a signature on that atmosphere because it would maintain the conditions far from thermodynamic equilibrium. The Life action on Earth continually pumps oxygen into the atmosphere, and without that action, the oxygen levels would decrease. It turns out that both Mars and Venus have over 95% of their atmosphere comprised of carbon dioxide with only trace levels of oxygen, water vapor, and other gases. Clearly, there are no Life forces at play that alter this composition—this composition is predicted by thermodynamic equilibrium. We are here assuming that this feature of Life-altered atmospheric composition is a general one, which we see to be a reasonable assumption. This is also a hypothesis increasingly testable as we discover more and more Earth-like planets. And, furthermore, the lifeless neighbors give some clue as to what the Earth's early atmosphere must have looked like.

THE OZONE LAYER

The increasing oxygen in the atmosphere had additional benefits for the development of complex life forms on Earth. Life forms originated and lived in water bodies, due to its special properties and ongoing need for water. But water also provided protection from solar ultraviolet (UV) radiation. On land, exposure to direct UV radiation would increase mutation and cell damage. Oxygen, once it became abundant in the lower atmosphere, could migrate to the stratosphere, dissociate driven by UV energy, and form ozone. Ozone effectively absorbs radiation in the $0.1-0.3 \, \mu m$ wavelength range at which most harmful UVB is transmitted. The ozone layer created by life-and-environment synergistic process formed around 600 million years ago, and it provided a protective shield, allowing aquatic organisms to colonize and spur biogenesis on land surfaces.

Without this regulatory ecosystem service, Life would be confined to aquatic and near aquatic environments (e.g., oceans, seas, estuaries, wetlands, mudflats, etc.). Continuing to the present day, the stratospheric ozone layer has been maintained by Life—environment processes. Damage to this layer by human industrial emissions—chlorofluorocarbons (CFCs)—destroyed ozone, but as the CFC emissions have decreased, the protective layer has begun to "heal," building back up toward its former thickness and size. During the period of decreased ozone due to CFC emissions, people have observed increased rates of skin cancer in humans (Slaper et al., 1996). And, increased UVB radiation has been implicated in damage to amphibians associated with widespread decline in this group of organisms that are susceptible to harm from UVB. While the physics and biology have been well understood, and people in science, policy, and industrial action arenas have worked in cooperation to diagnose the problem, prescribe change, and achieve remediation in this case, we believe the holistic implications of the origin and

maintenance of the ozone layer have not been fully understood or appreciated. Had we humans completely destroyed all ozone, for example, it seems unlikely that any technological fix or human actions alone would be sufficient to regenerate the ozone layer. This essential Earth feature is a unique product of the fully integrated Life—environment system operating in win—win mutualistic fashion.

SOILS

Soils are a third source of evidence that a win—win Life—environment relation is possible and normal. In forests, grasslands, and terrestrial environments, soils "spontaneously" (as a net, holistic, systemic result of Life—environment action) develop and increase in depth, structural complexity, fertility, and function in ways that support living communities existing on those soils (Van Breemen, 1993).

Similar to the steps of increasing complexity mentioned above in the case of oxygen, changes that result from integrated action of Life—environment build on each other and lead to successive increases in the quality of the environment in its capacity to support Life. The enhanced life forms and ecosystems stimulated as fertile soils develop act to increase the depth and fertility of soils even more, and a positive feedback loop is clearly operating (see Fig. 3.2 for depiction of this positive feedback between life and environment). Reports estimate upward of 1 billion bacterial cells and over 1000 species exist in a single gram of fertile soil.

The process of ecological succession—by which life forms colonize barren sites like new volcanic islands, or recolonize sites made barren by disturbance, and then develop over time—includes a key aspect linked to the development of soils on the site. The development of soils during succession greatly aids an increase in the biomass that the site can support (e.g., from lichens at the start to mature forests over hundreds to thousands of years) as well as the biodiversity on the site. Soils are also operative with the elemental, energy, and nutrient fluxes and the impact of Life and environment at other sites and at larger scales that a given site has during succession. For example, as soils develop on a site, that site plays an increasing role in regulation of the water, carbon, and nitrogen cycles (among others) and thus has greater impact on other sites downstream or downwind.

As Life actions lead to soil development, strong gradients are built, such as the decreasing concentrations of organic matter, carbon, and nitrogen with depth. In prior work (Fiscus, 2007), we found power law distributions of organic matter, carbon, and nitrogen in soils of Western Maryland that matched the gradients and distributions reported by others seen in soils worldwide (Jobbagy and Jackson, 2000, 2001). This far-from-equilibrium, highly nonrandom soil structure—likely a common phenomenon to anyone who has seen the vivid, striated color spectrum of a soil profile—is a very strong signature of Life—environment synergy. In addition to the changing color, soils show high levels of organic matter, carbon,

and nitrogen (and other essential life elements) near the surface with decreasing amounts going down to bedrock.

In part, based on this distinctive gradient structure, soils provide many functions aiding plant and animal life and the interdependent ecosystems in which they live. The long list of beneficial functions includes the abilities of soils to

1. absorb, store, and release water in ways that aid plant, animal, and microbial life locally;
2. play a role in the purification of water and the global hydrological cycle;
3. absorb, store, and release organic and inorganic nutrients aiding myriad life forms locally;
4. play significant roles in global carbon, nitrogen, and other elemental cycles;
5. recycle wastes and dead organisms and regenerate key inorganic nutrients via decomposers;
6. provide habitat for diverse biotic organisms; and
7. provide a physical basis for anchoring plants, particularly large trees.

Echoing many of these critical soil functions, the Soil Health Institute (SHI) states on its website (Soil Health Institute, 2017):

The concept of soil health is gaining widespread attention because it promotes agricultural practices that are not only good for the farmer, but also good for the environment. An abundance of research shows that improving soil health boosts crop yield, enhances water quality, increases drought resilience, reduces greenhouse gas emissions, increases carbon sequestration, provides pollinator habitat, and builds disease suppression.

SHI has developed a suite of indicators to define and measure soil health and states:

Soil health, like human health, is a complex and holistic concept. For example, when a person goes to a medical doctor, their health is not judged by blood pressure alone. Instead, many tests are used to assess their health. In a similar way, soil health is based on numerous chemical, physical, and biological measurements.

The SHI also works with multiple agriculture partners to develop and promote practices that restore, enhance, and maintain soil health. One could say that at the core, organic farming (OF) [or other restorative farming approaches; see, for example, "regenerative agriculture" as recently defined (The Carbon Underground, 2018)] is all about soil health. However, our reductionist approach has conditioned us to think of OF as what it is not, rather than what it is. A survey of students in introductory biology to the question of "What is OF?" produces a list that OF does not utilize biocides (herbicides, pesticides, insecticides, rodenticides, etc.), does not utilize synthetic fertilizers, does not utilize genetically modified organisms (GMOs), does not utilize growth hormones, does not use confined animal feeding operations, etc. A long list of NOTs, but never an

understanding of what OF is. In fact, the multitude of constraints sounds as though the farmer is putting herself at a disadvantage. In reality, the farmer recognizes that plants have evolved and adapted to survive and thrive; therefore, the aim is to create conditions where those foreign additions are not needed. This boils down to maintaining healthy soils.

This appreciation of how soil health is totally essential for human life, via agricultural provision of food and fiber and the ecosystem services listed above plus many more, is an excellent sign of well-placed understanding and value of a key Life—environment system. However, as for the ozone case above, we see that understanding and value of soil often stops at proximate values and drivers. The deeper implication of soils revealing and requiring a more holistic concept of life and environment as inseparably integrated is usually missed.

THE LIFE—ENVIRONMENT WIN—WIN IN SUMMARY

In each of the examples given above, we see that parts of the ecosphere that are normally thought of as abiotic—without life—are in fact heavily influenced and maintained by Life processes. Therefore, a better descriptor of such activities is *conbiotic* rather than abiotic to expressly acknowledge the role of Life in these environmental aspects (Fath, 2014; Fath and Mueller, 2018). In addition to the few forms of evidence here, it is interesting that Life is able to improve one primarily solid material form of its environment (soils, which admittedly have liquid and gaseous aspects as well), as well as two gas material forms of its environment (atmospheric O_2 and stratospheric O_3). We could add details of how Life actions serve to improve the aquatic and liquid forms of the environment in ways that likewise have reciprocal benefit for Life. The CO_2 and O_2 concentration of oceans, dissolved O_2 concentrations in freshwater, water purification, and resupply through the hydrological cycle, mineral concentrations in waters, and pH regulation are all influenced by integrated Life—environment actions, interactions, and ongoing interdependencies. Looking at all this evidence jointly, we see that Life (action of the Life—environment relation) has transformed and continually modifies all phases of matter—solid, gas, and liquid—of the planetary environment, in ways that aid and enhance the capacity of those materials to support Life.

We could go on—we could examine the generation and maintenance of habitat diversity (which in turn aids species diversity), creation of those deposits that became fossil fuels (starting with photosynthesis), and other aspects of biogeochemical cycles as strong and abundant evidence that a mutually beneficial Life—environment relation is not only possible but is plausibly the natural norm when considering the full history of Life.

THE LIFE—ENVIRONMENT INTERFACE

The paradigm we propose here, and seek collaborative assistance to refine, explains the fundamental antagonism between humans and environment, resolves the apparent contradiction between evolution driven by competition and the state of the world today, and indicates the way to successful mimicry of the net positive Life—environment relation. We not only go to great lengths to present this paradigm as so far developed, but also know well that it is still hypothetical—thus the need for collaboration and active participation to test, refine, and further develop it.

In order to attain what Life has achieved, we must first reimagine Life itself—instead of thinking of life and environment separately, we submit that we need to view the Life—environment system as a unified whole. One way to adopt this holistic approach is to see Life—environment as a single unified system, entity, and relationship, as indicated in the choice to draw the system boundary so as to integrate both life and its essential environmental context. We discuss this more below in diagram form, in view of work by Bernard Patten, and in later chapters in terms of lessons from systems ecology.

Matching the paradigm shift circumstances of Kuhn, our proposed paradigm redefines Life to integrate environmental processes, which we see then to provide a more realistic idea (i.e., better objective fit to empirical data, more balanced science with equal analysis and synthesis) of the fundamental nature of Life. This theory holds potential to succeed where prevailing reductionist and mechanistic theories are failing and provides us the potential for the mutually beneficial relationship between Life and environment.

This redefinition in general form is depicted in Fig. 3.2A and B. This alternative framing and systems model is an essential first step to associated and later steps we present below and in Chapter 4. For example, to achieve the potential win—win relationship, we assert the need for a new holistic, multi-model concept for Life that integrates two distinct and complementary life types, "discrete life" and "sustained life." For our full discussion of the discrete versus sustained life models, see Lesson 6 in Chapter 6. We also see the need to synthesize three unit models of Life now used but not fully integrated (1) the cell—organism—individual, (2) the community—ecosystem, and (3) the biosphere. For our full discussion of these three "holons" (Koestler, 1968) as interdependent and unfractionable Life unit models, see Principle 5 in Chapter 5.

For now, we compare and contrast the simple diagrams Fig. 3.2A and B and discuss their implications. We then look at a similar concept in Fig. 3.3, a diagram modified from Ulanowicz (1997) dealing with the "emergence" of whole system properties, or the lack thereof, again depending on the act of drawing the "arbitrary" (meaning intentional, decided, and chosen, as well as subjective) boundary of what is system, what is context (or environment), and what is the interface between.

In Fig. 3.2A, we see a generic system in which only Life components are defined as inside the system of study. We also see that the generalized environment is outside the chosen system boundary. The living ecological processes act out on the environmental/evolutionary stage which is not itself part of the system. This is similar to the isolated Venn diagram in Fig. 3.1A. The interface between the two separate systems is characterized by a positive impact of environment on

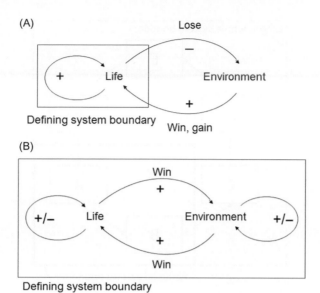

(A)

FIGURE 3.2

One key step to reframing our understanding of the Life—environment relation, and the ultimate net outcome of Life—environment interactions, is where we draw the system boundary. This in turn starts with the realization that drawing this boundary is a choice, and that this scientific choice has implications and ramifications that are profound.

Redrawn from Fiscus, D.A., Fath, B.D., Goerner, S., 2012. The tri-modal nature of life with implications for actualizing human-environmental sustainability. Emerg. Compl. Organ. 14 (3), 44—88.

Life (e.g., high-quality solar energy, rain water, oxygen, carbon dioxide, etc. coming in) and a negative impact of Life on environment (e.g., degraded and lower quality energy leaving, waste materials leaving, etc.). While this is an exaggerated and overly simplified cartoon, we see it as important to examine, and that the conceptual links to both physical systems and to human systems are potentially critical.

The alignment with physics is in reference to entropy—the generic physical system is understood to follow the Second Law of Thermodynamics, such that in any energy transformation, energy quality decreases and entropy increases. And for living systems, the general life system is assumed to be an open system that is able to operate and do work (metabolism, growth, cognition, and all the functions of life) only while receiving a steady input stream of high-quality energy, and while necessarily exporting degraded and lower quality energy. In the case of human systems, as said above, this generalized Life—environment (or system-environment) model may be employed (even if not consciously or explicitly) when we think, act, set policy, create technology, make plans, and govern ourselves with the assumption that degradation of the environment is to be expected, perhaps as a necessary evil or an unavoidable consequence of "doing business" as

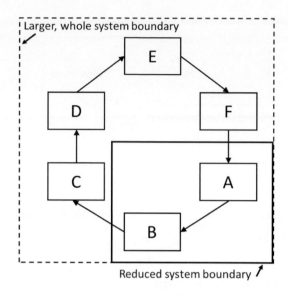

FIGURE 3.3

Modified from Ulanowicz (1997) whose original caption read "Two hierarchical views of an autocatalytic loop." The original boundary perspective (*solid line*) includes only part of the loop, which therefore appears to function quite mechanically. A broader vision encompasses the entire loop, and with it several nonmechanical attributes. Where Ulanowicz called the smaller system boundary the "Original system boundary," we have labeled it the "Reduced system boundary" associated with the reduced perspective and general approach of reductionism and analysis. We also added the letters to label each component.

we modern humans have come to assume is normal. A final note on Fig. 3.2A is that Life inside the system boundary is seen to impact itself positively, such as via reproduction, growth, development, evolution, etc.

In Fig. 3.2B, we see a very different generic system modeling approach. Now, the system boundary is chosen and drawn to include both Life and environment as a single integrated system. While some aspects of this are arbitrary and would benefit from further formalization (e.g., if, when, and why to label an impact arrow as $+$, $-$, or $\pm$; if we must always have an "outside" or external environment; if we must always depict high quality, energy coming in, etc.), the main feature of this alternative framing and system boundary is the ability to see a win—win relation between Life and environment. These positive impacts are related to those described above—Life improves the environment by way of soils, atmosphere, and more.

We develop several linked and similarly foundational ideas and definitions below in the sections on Holistic Perspective on the Origin of Life, and Holistic Perspective on "What is Life?", and we examine the complementary aspects of

Gaia theory below as well. Note also that these intentions and choices of drawing system boundaries for models and science relate strongly to the primary value emphasis on Life (Chapter 2).

There are important basic concepts with respect to modeling in science that we address in future chapters. For now, we offer just one quote and one reference as important for our use and strategies of models and modeling. Allen and Hoekstra (1992) provide the principle of imperfect modeling: "The past century, however, has taught us a little more humility concerning our ability to perceive reality directly. There always remains a veil that separates our models of the world from reality itself." And Rosen (1985) showed the multiple aspects which at best can lead to interdependence between our models and the real world in his modeling relation, where he has described the distinct processes of encoding, decoding, causality, and entailment (or inference).

We next examine an exercise compatible with Fig. 3.2 and with similar outcome, using a diagram modified from the one used by Ulanowicz (1997) as he explained the special properties of autocatalytic loops (see a fuller discussion of his work on autocatalysis below).

The main point of this diagram (Fig. 3.3) is again that our choice of the system boundary can have a profound impact on what we observe and how we are likely to interpret what we observe. Both of these are clearly related to our individual mindset, the current scientific paradigm and accepted practices of "normal science," and to our shared cultural mindset and everyday views and beliefs. For example, a common starting point is to study the direct relationship between one predator and one prey species, which could map on to components A and B in Fig. 3.3. As noted by Billings (1952), the breaking apart of an otherwise unified community (or ecosystem, or biosphere) is standard analytic practice, but, as we have noted, this is rarely followed by a counterpart process of the "synthesis of the total results" that Billings said was necessary. As the caption of Ulanowicz (1997) made clear, the choice of the reduced system boundary can align with the expectation, and a set of mathematical tools and physical assumptions, that causes and effects (or any of the predator−prey interactions), operate in linear fashion and can be studied effectively with mechanistic models. In the reduced system, it is clear that there is a zero-sum interaction with one winner and one loser. The larger, whole system boundary, in contrast, reveals a much more complex system that includes feedback and self-reference, and where mechanical behavior, valid fractionability of a system into simple components, and predominantly linear causality are now all highly questionable as starting assumptions, working hypotheses, and standard methods. It also raises the opportunity for win−win outcomes through the web of interactions.

These complex systems and philosophy of science themes and questions will continue to arise. We make a final comment and pose a "thought experiment" for now on Figs. 3.2 and 3.3. Consider the relative value and attractiveness of the smaller versus larger system boundaries in these two alternative science mission scenarios:

1. Your mission is to make an incremental advance in scientific understanding of life and/or environmental process and to produce a "pragmatic" (as based on social norms) contribution to science in the short term.
2. Your mission is to synthesize existing knowledge and to make an anticipatory contribution to science that has the potential to yield true human–environmental sustainability in the long term.

The ultimate mission context and the proximate tools of the trade shape each other. We have chosen mission #2 above, and thus the expanded and holistic system boundaries are more attractive. We next consider the contributions of another holistic scientist on the fundamental issues of systems, environments, and boundaries.

Bernard Patten contributed much work toward understanding our options for defining a system and its environment for the purposes of ecological modeling, scientific research, and understanding in fundamental ways how the world works. Perhaps the most basic and fundamental idea Patten provided is that we have two options for how to conceive environment, and how environment is conceived relative to any focal entity or living system of study. Patten (2001) explained the two options in context of human thought and science:

> In the history of human interactions with the external world, entities and environments became separated into two distinct categories, the first concrete and the second vague. This was due to a cognitive machinery that discerns objects but not, at least not directly, their covert linkages. Objects are local, whereas environments based on the transactions and relations between these are more extended, and boundaries which may be real or perceived separate the two. (p. 425–426)

And

> So it is natural for man that things and their environments are viewed as separate, and separated, and this entity–environment duality is registered strongly in physics' basic categories of open and nonisolated systems. The opposite, where environment and its defining entities are continuous and inseparable, is entity–environment synergy. (p. 426)

To understand these two conceptual choices—entity–environment duality and entity–environment synergy—is profoundly important, and Patten's work helps us know when and why each is appropriate and most beneficial to employ. One might argue that people readily conceptualize certain subsets of the environment as entities—"estuary," "meadow," "wetland," or "forest," for example, bring to mind well-defined regions, ecosystems, or local environments. The point is not that entities cannot be conceptualized, visualized, or even managed at various scales—the point is that for any subset of environment we may isolate in mind, model, and action there is always a larger, extended, background environment into which the subset is not only embedded but also fully interdependent for its continued existence. The object, in essence, emerges out of this context. Even larger scale entities such as "forest" or "estuary" must exist in mutual relation with a larger context and act in such way to maintain that context; otherwise, they would eventually destroy themselves by destroying their own context of existence. And, importantly, this concept applies to smaller entities such as "organism," including each human as an organism.

Patten (1982, 2001) credited Uexküll (1926) and his idea of function-circles (among other ideas) for the inspiration for his own concepts of environment. From the starting point of these two categorical options—duality versus synergy (or unity)—Patten (1978, 1982) built his own theory and methods based on environs. Following Uexküll as well as Gibson, Patten defined environs as relativistic—they are constructed relative to focal entities or components within a network or system model. That is, a distinct environ exists for each species or functional group within an ecosystem model and is defined in specific reference to each species or component.

Patten (2001) explained the major characteristics of environs—how they are defined and used. Perhaps most importantly, each system component has separate and distinct input and output environs. Additional key characteristics include (1) environs are computable by the input—output methods derived from Leontief and Hannon, (2) taken together they account for all the matter and energy in a modeled system (they are exhaustive), (3) no environ shares the energy-matter of another (they are mutually exclusive) (these two previous characteristics give the environ the formal description of a partition), (4) they "reflect transactive causality" (that based on direct exchanges of matter and energy) and "encompass relational causality" (indirect causes and effects between components), and (5) no two environs are alike qualitatively or quantitatively. As reported in Fath and Patten (1999) network environ methods are also unique in the use of matrix power series during analysis. Fig. 3.4 compares the dualism and synergy views and shows separate input and output environs in Patten's synergy view. In this

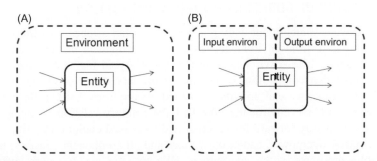

FIGURE 3.4

Modified from Fath and Patten (1999). (A) Entity—environment dualism, the mainstream paradigm. (B) Entity—environment synergism and distinct input and output environs, Patten's paradigm.

case, the object is truly inseparable from its environment because it comprises both an input environ and an output environ.

Patten (1982) traced the history related to these ideas back to Darwin:

> Before Darwin (1859) environment was considered an organic whole. Everything in it made some contribution and had some meaning with respect to everything else. Darwin subscribed to this view, but his emphasis, and that of his followers, on the evolving organism struggling to survive, suppressed the exploration of holistic aspects of the origin of species that might have been developed. After Darwin, the organism came into great focus, first as a comparative anatomical entity, then later with physiological, cellular, molecular, behavioral, and genetic detail. In contrast, the organism's environment blurred through relative inattention into a fuzzy generality. The result was two distinct things (dualism), organism and environment, supplanting the original organism-environment whole (synergism). (p. 179)

One approach to the choice posed by Patten is that we could use both synergy and dualism methods and compare the results. This shares a basic strategy of "dialectical thinking" as described and promoted by Elbow (1986), and which we discuss in depth in Chapter 7. For now, we will work with the view that entity—environment dualism is the fundamental approach of the current mainstream scientific paradigm. As such, we suggest the need to examine this fundamental approach to understand how the world works, and we are exploring whether this very early step in scientific process—drawing the boundary between a system of study and its environmental context—may also be partly responsible for our current chronic and systemic environmental crisis. Next, we merge two prior discussions—the primary role of competition as explaining change and improvement (increased fitness, adaptation to the environment) over time, and this boundary or interface between life and environment.

THE SYNERGY OF COOPERATION AND COMPETITION

As long as we assume, in our daily practices of normal science, that life is fundamentally separate from environment (as in Patten's dualism in Fig. 3.4A, and as in the fractioned Life—environment system boundary in Fig. 3.2A), and as long as we assume that competition is the sole or primary driver of change in living systems over time, we end up with the added assumptions that (1) life can be in competition with the environment (or that the primary relation is antagonistic) and (2) the best way forward for progress and increased evolutionary fitness is to defeat, control, overpower, and/or outwit any and all components, factors, threats, or variables presented by the environment. [We note the need to clearly differentiate between transaction or predation, which is the reduced system of Fig. 3.3 and competition, the focus here. Transaction or predation occurs pairwise between two components (species, organisms, etc.) but for competition to be realized, there must be a third actor or resource over which to compete, since the two competitors have no direct transactional connection. Given this differentiation, we see the general conceptual point to be valid.]

As we have suggested, this perspective can also lead to the sequence of linked causes and effects in Fig. 1.1 which emanate outward from the center of our science-based system of ideas (restating with a few modifications):

A. In the science paradigm, life is separated from environment, thus severing the unity of life and life support systems conceptually and scientifically.

B. Following the scientific paradigm, life is separated from environment in mind and action in the wider cultural system of ideas. Note that culture can have a reciprocal and reinforcing role to help continue the prevailing science paradigm as industry rewards reductionist science, for example.

C. Once fragmented, it is possible and likely that the value of environment is seen and treated as less than the value of life.

D. Individuals (as well as many nations, corporations, organizations, and other social entities) act for self-interest primarily and compete for what they perceive as limited, scarce, and zero-sum resources and assume that environment degradation is normal, expected, inevitable, and acceptable.

E. Environment is consumed and degraded as manifest in many symptoms of ecological crisis, and the influence of the citizens' mental fragmentation and devaluation of environment travels upward to larger scales and produces the global crisis.

We also noted that it is not possible for this relative devaluation of environment (and fundamentally antagonistic perspective) to occur if Life−environment remains unified as a single focal entity and system of study. Either the combined unit improves together or declines together. And, we have shown that in nonhuman living systems, a positive impact of life on environment is readily observed in the ways that atmosphere, soils, and other crucial aspects of the environment improve over time. Thus, we are in search of an alternative approach by which our science paradigm, and the linked cultural system of ideas, can help us to understand and then achieve this same positive relation with environment. We next look at work of Ulanowicz on his alternative explanation for how evolution and competition function. We then seek to extend his model to integrate nonliving components of the environment as inspired by the synergy model of Patten.

As mentioned before, Robert Ulanowicz has led the way in developing a new holistic paradigm for science based on ecology that we seek to support and build on here. He has contributed a large body of work in hundreds of papers and several key books. In this section, we focus mainly on his radical and alternative explanation for the first and most fundamental basis for evolution, competition, and selection. We note along the way several of his other revolutionary ideas and concepts that are inseparably linked to his model for evolution.

In his most recent book (Ulanowicz, 2009a,b), he provided the "axioms for life" which also serve as the central tenets of process ecology and of his ecological metaphysic. While already paraphrased in Chapter 2, we review and quote these here, as they are relevant and applicable for the central set of concepts—how autocatalysis and indirect mutualism provide a new perspective on evolution and self-organization in living systems.

His book, *A Third Window* (Ulanowicz, 2009b), contained three fundamental postulates for an ecology that Ulanowicz also described (quoting Hutchinson) as "the study of the Universe":

1. The operation of any system is vulnerable to disruption by chance events.
2. A process, via mediation by other processes, may be capable of influencing itself.
3. Systems differ from one another according to their history, some of which is recorded in their material configurations.

These appear deceptively concise, perhaps even incomplete or insufficient to explain the basic processes and dynamics of Life's unique properties and abilities for growth, development, evolution, and adaptation, but we will see that they provide a very sturdy foundation on which to build a full explanatory paradigm.

In his prior book, *Ecology: The Ascendent Perspective* (Ulanowicz, 1997; we abbreviate this book as EAP), Ulanowicz presented his explanation for how autocatalytic loops and indirect mutualism, along with the expanded view and holistic modeling that allows these features to emerge, to become visible (see Fig. 3.3), tell a very different story than the mainstream mechanistic view of life and evolutionary change. We paraphrase the main thread of his ideas here (Ulanowicz, 1997).

Seeking a general and systems explanation for order, ordering, organization, development, agency, formal and final causes, Ulanowicz considered the nine possible pairwise interactions between any two processes that may come into contact. Each can have on the other an impact that is positive or beneficial, denoted by $+$, negative or detrimental $-$, or neutral or no impact, 0.

He then identified as "mutualism" the doubly beneficial interaction $(+, +)$ in which each process aids the other. He noted that this pairwise interaction is unique among the nine possible combinations, stating that mutualism "leads to qualitatively different behavior" than any of the others, and it generates "nonmechanical behaviors."

Ulanowicz next extended this examination to sets of more than two interacting processes, beyond pairwise interactions to the next simplest configuration—a three node (or component) network (or system). In the case where all three individual interactions around the three-node loop are positive $(+, +, +)$, Ulanowicz noted how this leads to "indirect mutualism." He further noted that the property of indirect mutualism can apply to networks/systems of any size (any number of nodes or components).

The language, modeling, and terminology Ulanowicz used are clear and precise. Still, it is interesting to note several subtle variations in his expressions and explanations. Throughout, he emphasized that the relations, interactions, and influences between any two processes, nodes, or components are flexible, variable, and never fixed or fully determined. Borrowing from Popper, he used the term "propensity" to describe the general tendency, or overall effect, of any interaction or influence. So, an arrow with a + sign showing a positive interaction from component A to component B means that component A has a propensity to increase the rate of B. Imagining that not all instances of the influence of A on B are the same, and not all are positive, the + can mean that the propensity for a beneficial influence is greater overall than for a detrimental influence. In addition to these usages, Ulanowicz also said that A will increase the activity of B. All of these terms—positive, beneficial, to increase the rate, to increase the activity— make sense, and all are related and understandable when considering a general set of processes such as those energy and matter transformations and reactions of chemistry, biology, ecology, and the environment.

Following the influences around the entire three-node loop (+, +, +), Ulanowicz noted crucially that component A can influence itself. Note, too, that this principle ranks in the top three of his axioms of life. Thus, the three-node loop of indirect mutualism is "autocatalytic," a term meaning self-enhancing. This is an important concept Ulanowicz gleaned from his early academic study and PhD training as a chemical engineer.

In EAP (Ulanowicz, 1997), Ulanowicz noted that autocatalytic systems are known and treated as *mechanisms* in chemistry, as all the reactants (similar to components as we have called them here) are fixed. The chemical reactants do not themselves change, and their modes and specifics of reaction (process) do not change, and so all dynamics and interactions are strictly determined and predictable—two signatures of mechanical behavior. But Ulanowicz stated that this is not the case in ecology and life science, where the participants or components *can and do* change, as they are more complex and adaptable. This makes all the difference, and Ulanowicz asserted that in the ecological case, the autocatalytic loop system is profoundly different—it is *nonmechanical*.

To help ground these abstract concepts in a realized and functioning system, Ulanowicz presented a case study and real biological/ecological example and showed how a three-node system with *Utricularia*, periphyton, and zooplankton maps onto his generalized three-node autocatalytic system. The bladderwort *Utricularia* aids the periphyton by providing it a substrate to grow on, the periphyton is food for the zooplankton, and the zooplankton becomes food for the bladderwort when sucked into its utricle trap, thus completing the (+, +, +) influences around the three-node self-enhancing loop. This is a clear and accessible example, but by no means a special case. Any time we construct a food web or network of feeding relations, we see that many loops exist; thus, indirect mutualism and autocatalytic loops are ubiquitous in all ecosystems. For other examples, consider the very large set of food web network cases in which energy or material passes from top predators through soil and decomposers and thereby feeds and aids components (species, functional groups, plants, microbes, etc.) at or near the base of the food chain.

Ulanowicz next described eight nonmechanical properties of this special form of self-enhancing self-organizing life system. All eight of the special properties are interrelated, and the cumulative effect can be described in many ways. Regarding perhaps the most striking one of these, Ulanowicz (1997) wrote:

> In an autocatalytic system, *an increase in the activity of any participant will tend to increase the activities of all the others as well.* (p. 42)

This is a noteworthy special property—we could say that this is the locus of a special case of the part—whole relation of Life. This is in essence a statement that the parts aid the whole, each part aids each other part, and the whole serves to aid each part. This kind of multi-scale unity and organic coherence is familiar in how all organs and subsystems of an organism serve to aid all others and the organism as a whole, and how all functional groups serve to aid all others and the ecosystem as a whole with ecosystems.

Interestingly, systems thinker, architect, and urban pattern guru Christopher Alexander made a similar observation about how buildings and other aspects of the built environment relate with each other: "Each center is (recursively) dependent on other coherent centers for its own coherence" (Alexander, 2012, p. 428). In his terminology, a "center" is a quasi-autonomous building that has a defined function and purpose. It is coherent in the sense that it performs this function as a bank or hotel or post office. But it is dependent for its continuation on the other buildings around it, which are each performing their own coherent functions. A thriving community needs each of these centers interacting with each other. The quality and health of the entire community emerges out of these contextual relations. This demonstrates the foundational role of the concept of positive, feed-forward relations in socioeconomic systems as well as ecological ones. We expect that development of a more pervasive holistic science will spawn examples in many fields.

It is difficult to paraphrase EAP (Ulanowicz, 1997) which is already concise; this recapitulation does not do justice to the original, and we highly recommend that everyone read EAP. But we borrow and share these bullet point ideas here, as they are essential to the new holistic science we seek to promote. Ulanowicz described each of the eight special properties of autocatalytic systems. It may help to refer to Fig. 3.3 for reference. He explained how autocatalysis induces or embodies the following system qualities (with our versions of his descriptions of each):

1. Growth-enhancing—an increase in the activity (or rate, or material flow) of any component leads to greater activity in all other components and in the network as a whole. The configuration also results in greater activity of the whole than the activity if the components were operating separately.

2. Selection pressure—random changes in any component's behavior propagate around the loop to either reward changes that increase the mutualism (either increase its sensitivity to a prior component or increase its catalytic benefit to a following component) or to penalize changes that decrease the mutualism.

3. Asymmetry—as Ulanowicz wrote (1997), "Unlike Newtonian forces, which always act in equal and opposite directions, the selection pressure associated with autocatalysis is inherently *asymmetric*." The propagation of positive incremental changes forward, and their reward and amplification when the effect returns to the originating component, impart a directionality to the operation of the looped network as a whole, thus tending to ratchet up activity, rate, and size of the system as a whole.

4. Centripetality—the autocatalytic system operates so as to pull in material and useable energy (or exergy) from the environment. This follows from the same kinds of reward loop impact as above—any modification to a component that leads to increased input of energy or material into that component, and that then enhances activity around the loop, will come full circle to reward that increased input modification, and so on.

5. Competition—the autocatalytic system configuration induces competition not only among varying properties of components (as under #2 "selection pressure") but also between components themselves. This occurs since more than one pathway, with alternatives for any component as part of a pathway, can exist. If some new component arises and begins to participate in the loop, the mutualism and reward loop will act to penalize and eventually replace less sensitive or less catalytic components where the flow can follow the better pathway.

6. Autonomous—Ulanowicz described several ways in which autocatalytic systems are (to varying degrees) independent of their elemental constitution and environmental context. He described one aspect of this due to centripetality by which the system "actively creates its own domain of influence." He also noted that this "creative behavior imparts a separate identity and ontological status to the configuration." Finally, he described how "the characteristic time (duration) of the larger autocatalytic form is longer than that of its constituents" leading to its "persistence of form beyond present makeup." Just as cells in organisms turn over and are replaced while the organism remains, elements within cells are replaced more rapidly still, and species come and go while ecosystems maintain their essential form and function, such autonomy and persistence of form is a familiar feature of living systems. In centripetality and the autocatalytic loop, Ulanowicz saw the "most primitive hint of entification, selfhood, and id."

7. Emergence—the autocatalytic configuration, as well as the other seven associated special, nonmechanical properties, "emerges whenever the scale of observation becomes large enough." Again, see Fig. 3.3 for a graphical depiction of how the picture changes from a mechanical and simple, linear cause—effect system, to the complex, holistic, and organic autocatalytic loop system.

8. Represents a formal cause in the sense of Aristotle—whereas most explanations in normal science refer to material or efficient causes (often considered mechanisms) operating at a level below the focal level (the level at which the system is defined, and its boundary is set, and observations, experiments, and models are made), or occasionally to larger scale causes

operating at a level above the focal level, with autocatalysis "agency [much like causality] can arise quite naturally at the very level of observation." Ulanowicz went on to say that autocatalysis not only "takes on the guise of a *formal* cause, *sensu* Aristotle," but also that the directionality and asymmetry in autocatalytic loops represent "telos ... a very local manifestation of final cause ...." For Ulanowicz to associate autocatalytic loops with formal and final causes is to us a major indication that this organizational form and configuration of processes is highly unique, critical to understand Life, and with potential to help solve the systemic problems we face with both our science paradigm and our impacts on our Life support systems.

Taken together, the eight nonmechanical properties of autocatalysis form the theoretical basis by which Ulanowicz has shown that mutuality is plausibly prior to, and the generating source of, competition and evolution. These sections of his later book add more emphasis to this new paradigm (Ulanowicz, 2009b):

> ... in connection with Darwin's theory, a very important but unstated premise of his scenario is that participants strive to capture and accumulate resources. The conventional Darwinian narrative does not mention the origins of this drive, but we now see it is as the deductive consequence of autocatalytic action. (p. 72)

And

> To underscore the fundamental and essential status [of] centripetality, we now assert that competition is derivative by comparison. That is, whenever two or more autocatalytic loops draw from the same pool of resources, it is their autocatalytic centripetality that *induces competition* between them. By way of example, we notice that, whenever two loops partially overlap, the outcome could be the exclusion of one of the loops. (p. 73)

And

> One should never lose sight of the fact that the autocatalytic scheme is predicated on mutual beneficence, or more simply put, upon mutuality .... That competition derives from mutuality and not vice versa represents an important inversion in the ontology of actions. The new ordination helps to clear up some matters. For example, competition has been absolutely central to Darwinian evolution, and that heavy emphasis has rendered the origins of cooperation and altruism obscure, at best ... these efforts ... invariably misplace mutuality in the scheme of things. Properly seen, it is the platform from which competition can launch: without mutuality at some lower level, competition at higher levels simply cannot occur. (p. 75)

As we have said, such positive feedback, autocatalytic loops are ubiquitous in cycles of energy and materials in food webs. The work of Ulanowicz on this unique community-ecosystem level property provides better understanding of evolution, natural selection, competition, and the striving to self-improvement of living systems. This systemic striving, as Ulanowicz suggested, is missing from mainstream Darwinian-theory based on unit models of organisms, individuals, and species, none of which are modeled or defined as having innate positive feedback or autocatalytic loops.

Even with this strong emphasis on the primary explanatory and ontological role of mutualism, autocatalysis, and centripetality, we might say that Ulanowicz has described various forms of synergy between competition and cooperation and indicated ways in which they overlap and are interdependent. Whether seen as primary, or as one half of a dialectic, this view of mutualism is a qualitatively different approach than the current mainstream life science paradigm. See Chapter 6, Lesson 5, for more on the fundamental role of mutualism as shown by systems and network ecology. Overall, we assert that this perspective is a critical pillar of a new science paradigm capable of helping us see and then achieve a mutually beneficial relation of humans and environment.

The next step is to extend the work of Ulanowicz on autocatalytic loops and indirect mutualism to incorporate an active, participatory, and self-transforming role of the environment. In the diagrams, models, scenarios, and examples he used, Ulanowicz depicted the components or participants in autocatalytic loops primarily with reference to living components—species or functional groups, as in his *Utricularia* case. While this is enough to shift radically the paradigm of Life science and to elucidate a new and wholly missing fundamental concept to evolutionary theory, we can do even more if we apply this approach while including environmental components in the loops.

Since we have described already how soils and atmosphere improve over time in terms of their quality and capacity to aid Life, we can start by imagining one of these environmental factors as an active component inside an autocatalytic loop that also includes living components. See Fig. 3.5 for a three-node loop example.

Once we have an environmental process in the loop, then all of the eight non-mechanical properties of autocatalysis (Ulanowicz, 1997) extend to the integrated environment. For example, an increase in the rate or activity of the environmental process will increase the rate/activity of the others—the living components—and vice versa. A change in the activity of one of the Life processes will alter the activity of the environmental process. This straight-forward extension of Ulanowicz' autocatalysis is the seed germ of the win−win Life−environment relation and helps to understand Life−environment as an interdependent, unified, and co-evolving system. This model also helps explain the strong evidence of improvement in soils (e.g., how over time they grow in depth, increase in fertility, and build nonrandom gradients in key constituents like organic matter, carbon, and nitrogen). And, if we replace soil with atmosphere in Fig. 3.5 as an active environmental component, then this model helps explain improvement in the atmosphere (e.g., how over time the oxygen content aided the diversity and energy processing capacity of Life, and how the ozone layer protected against UV damage). This conceptual model would also apply to the seed germ of Lovelock's Gaia theory of a living biosphere.

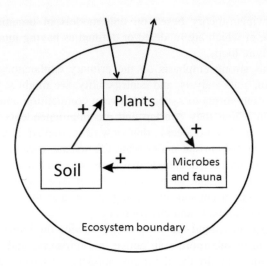

FIGURE 3.5

Modification of the three-node autocatalytic network of Ulanowicz (1997) to include an environmental compartment such as soil shows how mutually beneficial relations can integrate environmental roles and lead to enhancement of life support capacity. We could draw a similar diagram with atmosphere, surface waters, or other environmental components in place of soil.

The plot thickens as we examine details of some of the environmental processes that can and do participate in autocatalytic loops in close mutualism with living components. In order for the impact or result of a component such as soil to have a beneficial (+ sign in the diagram) impact on the next component around the loop, we must be able to observe a nonliving process, or perhaps a hybrid living and nonliving process, that enhances the next process downstream. With respect to the origination of this loop, as we discuss later related to the origin of Life, this is akin to saying that some natural, spontaneous tendency explainable by physics and chemistry becomes an asset to the loop as a whole, is able to be augmented by the component feeding it, and to augment the component that it feeds. And on the flip side, this suggests that living component processes will operate in ways that increase the rate of activity of *certain* nonliving processes—those that can enhance the next component in the loop and participate in the autocatalytic system as a whole. The whole set of interactions becomes a pay-it-forward complex, which, however, is unintentional in the proximate sense—there is no designer—but closed in the extended sense that those that contribute to the autocatalytic processes best promote existence of the overall system. As systems-thinking urbanist Jane Jacobs once remarked: "It may be that all self-sustaining systems are reciprocating" (Jacobs, 1969, p. 126).

One way to think of this beneficial physical/chemical process is in terms of a spontaneous environmental capacity to break down structures, and to dissipate energy gradients, as required by the Second Law of Thermodynamics. Ulanowicz (2009b) attributed such a dissipative capacity to autocatalytic loops in general:

> ... as autocatalysis drives up the aggregate levels of system activity, it concomitantly inflates the rate of system dissipation. (p. 94)

This overall system tendency to increase dissipation is compatible with an integral role for a component that breaks down chemical bonds that may have been formed, that "decomposes" molecular strings that may have been assembled or "composed," by an earlier process step in the loop. By analogy with Fig. 3.5, this would fit with a process by which organic forms of nitrogen are mineralized back to inorganic forms (NH_4 or NH_3), and this would be a positive process step to increase the rate by which a plant component (or even a proto-plant, or proto-autotroph, like a general "composer") would be able to take in nitrogen and grow, thereby feeding the next component around the loop. A similar impact occurred as increased oxygen concentration in the atmosphere increased the redox potential of the environment, thus favoring greater energy-dissipating reactions.

The idea of a molecular string "composer" as a generalized proto-autotrophic functional process operating in mutualism with a molecular string "decomposer" as a generalized proto-heterotrophic process is something we presented in prior work (Fiscus, 2001–2002). We use the term composer here, instead of the usual term "producer" but with essentially the same meaning, to emphasize the complementarity of function with the decomposer. We discuss this more in reference to the origin of Life but note here that the systems concept of "coupled transformers" is one that borrows from Alfred Lotka (1925) and one that is in harmony with the "two-tendency universe" as Ulanowicz (1997) described the ubiquitous synergy of natural processes that build up coexisting with processes that tear down. This quote of Mikulecky (1999) summarizes this idea well (where he was referring to work of Robert Rosen and Rosen's holistic metabolism-repair model of organismal life):

> One of the first and most crucial aspects of the evolving living system was its failure to last! It was in a condition of being torn down as fast as it was being built up and this is what allowed it to evolve. Stability is the return to a condition after being perturbed from it. How much more stable could something be than to have both its construction and destruction under strict limits? Both construction and destruction are systems properties. The systematic tearing down allows rebuilding, replication, and evolution.

If we replace the soil component with an environmental atmosphere component, then a similar scenario could be envisioned with reference to organic carbon compounds—one living (or proto-living) functional component forms organic carbon compounds, forming bonds and assembling "useful" structure, and another component (perhaps originally as fully abiotic) naturally breaks those carbon bonds, dissipates the energy, and returns original building blocks to the next component, perhaps in the form of CO_2 as an input.

This integral Life role for a decomposer, dissipation, or deconstruction function hints at some profound issues. In essence, this view of the original and fundamental nature of Life—environment systems and how they are organized and operate—where we start with an ecological or ecosystemic unit model in mind rather than a biological or organismal one—suggests that *death and life co-arose and have always coevolved*. This echoes the question that Patten (2014, 2016a,b) has posed and has studied in depth:

> Is life's destruction of current life (eating and being eaten) to sustain and reconfigure new life to fit a changing planet the only way to organize a biosphere?

As we continue to explore Patten's rich question (first mentioned in Chapter 1) and the concept of Life—environment as integrated whole, we may find that another principle is a logical extension—death and life must always go together and coevolve, as a form of creation and renewal.

In spite of a human longing for everlasting life, dissipative tendencies are unavoidable in a thermodynamically lawful world. The necessity for new energy and the processing that it ensues to maintain the complex, self-organized structures wears down the very system it is nourishing. The design strategy of nature has not been to extend this timeline indefinitely, but rather one stays long enough to leave one's mark that this process of birth and death can function in perpetuity in a broad cycle of life and death, production and consumption, composition and decomposition, creation and renewal.

The strategic integration of construction and deconstruction is also a useful technological approach to sustainable materials use, as in the *Cradle to Cradle* design philosophy of William McDonough, Michael Braungart, and associates (McDonough and M. Braungart, 2002; also see: www.c2ccertified.org/ for current certification and licensing process). We examine this inherently cyclical application to manufacturing and use, which we might call "ecomimicry," in Chapter 8.

Tying back to our discussion on soils, Wendell Berry eloquently remarked, "Soil is the great connector of lives, the source and destination of all …. It is alive itself. It is a grave, too, of course." He was writing about the current agricultural crisis and the forcing of industrial chains rather than ecological cycles and concluded that it is essential that we "Establish agriculture upon the same unifying cycle that preserves health, fertility, and renewal in nature by which 'Death supersedes life and life rises again from what is dead and decayed'" (Berry, 1977, p. 90).

In this chapter, we have looked at our choices related to drawing system boundaries, how those choices impact the results and the values/assumptions promulgated, options for conceptualizing the entity—environment relationship, the unique power of autocatalysis to explain Life in a way that also suggests a new view of evolution, the synergy of cooperation and competition, and the synergy of life and death. We next apply the holistic and anticipatory approach to study the origin of Life and to address the question "What is Life?".

Life: From origins to humans

4

In the last chapter, we looked at our choices related to drawing system boundaries, how those choices impact the results we get, and how values and assumptions play into our Life science. With these strategies and framing in mind, we next apply the holistic and anticipatory approach to study the origin of Life. After that, we review existing approaches to answering the question "What is Life?" and offer our holistic contribution to this recurring and provocative question.

HOLISTIC PERSPECTIVE ON THE ORIGIN OF LIFE

How does the holistic Life perspective alter attempts to develop hypotheses, models, scenarios, and narratives for the origin of Life? This is an important question, not only due to the perennial interest in the question, and the potent curiosity and basic science involved, but also due to the possibility that understanding the original and fundamental nature of Life may hold clues and keys to what we need to understand and change to live long and prosper as technologically advanced humans in socially developed nations and civilization. If that link—between the origin of Life and implications for actionable science to steer human progress—seems weak, then we hope you will read this next section with an open mind and consider that it may yield important fruits. This quote may serve for inspiration—a motto that at one time was posted on the Internet as the stated mission of the NASA Institute for Advanced Concepts (NIAC). The mission of NIAC was:

> To understand life from origin to destiny

If memory serves, this mission was on the NIAC website circa 1999. The group is now called NASA Innovative Advanced Concepts (again, NIAC) Program, and this holistic mission quote is no longer posted on their websites, as far as we can tell. It remains an inspiration for us, and even if only temporarily a way of thinking for NIAC, it serves to concisely encapsulate the idea that Life's origin and

Foundations for Sustainability. DOI: https://doi.org/10.1016/B978-0-12-811460-5.00004-2

open-ended future are necessarily linked. NASA's Office of Space Science does currently list a similar mission, but with reference to the universe as a whole (NASA OSS, 2017):

> The mission of the Space Science Enterprise is to solve mysteries of the universe, explore the solar system, discover planets around other stars, search for life beyond Earth; from origins to destiny, chart the evolution of the universe and understand its galaxies, stars, planets, and life.

The NASA Astrobiology group is similar, defining astrobiology as "the study of the origin, evolution, distribution, and future of life in the universe" (NASA Astrobiology, 2017).

As we mentioned in Chapter 1, we also see this original/fundamental and destiny/sustainability unity, and that holistic science and an ecosystemic perspective can help, when we compare the challenges of sustaining Life on Earth and successfully colonizing Life beyond Earth. These two pursuits are interestingly entangled, at least by our way of speculating: ironically, it may be that we cannot know and achieve sustainability of Life on Earth until we know and achieve colonization of Life beyond Earth, and vice versa. It was the photos of Earth from space, e.g., the 1972 *The Blue Marble*, that galvanized the environmental movement's focus on a global perspective, and ideas such as Boulding's "Spaceship Earth" that made it clear we are dealing with similar life-support issues as our astronauts, albeit on a different scale. This same paradoxical consideration applies to the origins and destiny extremes as well: we may not be able to know and achieve sustainability of Life on Earth until we understand well the original and fundamental nature of Life.

In prior work, Fiscus (2001-2002) proposed an "ecosystemic life hypothesis" that was compatible with the ecological origin of life scenario of Odum (1971). In this hypothesis, similar to the discussion above, Life arose as an integrated set of coupled complementary processes of molecular string "composers" (analogous to proto-autotrophic processes) and molecular string "decomposers" (analogous to proto-heterotrophic processes or organisms). Odum (1971) described two similar coupled chemical reactions as representing prebiotic "production" and "respiration" and depicted them as arising prior to cells and organisms as aided by hydrological cycles, thermodynamic gradients, and photochemical reactions in "circulating seas." Odum's scenario also depicted this ecological cycle as later generating cells and organisms via "encapsulation and miniaturization."

Instead of a solely bottom-up approach—the origin of "cellular life" (Morowitz, 1992) as most mainstream scientists have imagined it—in new paradigm we offer hinges on a multi-scale perspective. We hypothesize that the biospheric, whole-Earth aspect of Life played a key role, most likely as related to the hydrological cycle. The backbone of the planetary water cycle is an interesting prototype for Life as a general dynamic system: two main phase transitions of water (evaporation and precipitation), and the dialectical interplay of solar radiation (driving evaporation) and Earth's gravitation (driving precipitation), form a cycle that sustains itself indefinitely.

Another feature of the hydrological cycle we have explored as potentially implicated in the origin of Life is the interface between freshwater runoff, estuarine dynamics, and ocean circulation. A region of unique dynamics, where the phenomenon called the estuarine turbidity maximum (ETM) occurs, displays complex characteristics at the place where freshwater from land (river input) meets salt water coming in the from the ocean (aided by estuarine and tidal flows). A well-studied example of this ETM zone exists in the Chesapeake Bay (North and Houde, 2001). The physical and chemical interactions create a stratification and boundary layer between the salt and fresh waters. The ETM zone also has a high degree of mixing, circulation, and resuspension of particles, playing roles much like Odum's (1971) "circulating seas." These and other features make it a prime candidate for the location of a systemic origin of Life at the intersection of planetary, regional, local, and microscopic dynamics, and creative forces and elements involving water, land, atmosphere, and ocean. This scenario also integrates the three Life modes, and three Life unit models, we have employed to explain Life—the biosphere (or planet), the ecosystem (as in the proto-plant and proto-animal functions), and the cell/organism (microscopic and discrete life entities).

An existing Life feature that may be a legacy of an origin at the ETM is the common combination of (1) an internally salty cellular and physiological make-up with (2) the continual need for freshwater input observed in most life forms. A literature survey would be useful to examine how common this is among all life forms. A related issue worthy of study is the timing of ocean/estuarine salt content, the origin of Life and important developmental stages of early Life. Are the time scales compatible? How did ocean salinity change in relation to dynamics and processes in the prebiotic world and into the protobiotic world? These questions are candidates for follow-up work based on our theory, and they would help to confirm or refute the holistic Life—environment hypothesis.

Many other properties of water are important for understanding and aiding Life, but we do not address them in depth here. Properties such as water's specific heat capacity, temperature—density relation, ion solubility, hydrophilic and phobic membranes, absorption of ultraviolet radiation (UVB), and many others, while key to Life, are not essential for explaining and differentiating the ideas and paradigm we present here. Thus, they are explained well in many other works and references.

Another approach, an alternative to starting with the physical/chemical dynamics at the ETM, or with molecular string composers—decomposers in proto-symbiosis, is a line of "relational reasoning" that may amplify the value of a holistic and ecosystemic origin of Life scenario. Next, we consider the origin of Life in relational terms and see support for our idea that the primary Life—environment relation likely has been a win—win relation from the origin onward.

Imagine a pre-Life Earth environment characterized by an undifferentiated wholeness. In this hypothetical era existed no life or Life, no living cells, organisms, ecosystems, or biosphere, and no living "self" existed as distinct from environment. In this time and space, in the most general sense, we could imagine that only a single "relation" existed: the environment—environment relation (see below for features of "relation" as we are using the term here). The logic here is that the other key generic relation we seek to understand and heal (to achieve human−environmental sustainability)—the Life−environment relation—did not yet exist. Starting from this original whole and unitary environment−environment relation, we then seek to understand the origin of Life, and the origin of the Life−environment relationship, in broad conceptual terms, and importantly, in relational terms.

To add more details to the simplest pre-Life situation, we imagine two aspects of environment as nonliving processes or objects characterized as energy and/or material. The link between these two processes/objects is a relation of some sort, such as a transformation (e.g., liquid water evaporates to water vapor), interaction (e.g., two molecules of water collide), or some other form of material, physical, chemical, energy, gravitational, etc. influence or effect. From this starting premise, we suggest the next set of logical steps, eight propositions and a working hypothesis. The propositions are as follows (modified slightly from Fiscus, 2013):

1. Relations are *not* material. Relations *are* physical and real.
2. Relations are not localized and are not objects. Relations are non-localized and exist between objects or processes. Relations can be transformations, interactions, forms of organization, configurations, or influences. They are not conserved and can multiply combinatorically.
3. The origin of Life involved "something new" arising from the pre-Life realm of environment and relations.
4. The "something new" that arose was not material, energy, or any kind of object or "stuff." [We note that this is compatible with the first law of thermodynamics—matter-energy is not created or destroyed; thus, matter and energy are conserved at the origin of Life.]
5. The "something new" that arose at the origin of Life was a novel emergence in the realm of relations. For example, something new could have arisen in the organization of matter-energy, or in the configuration of some set of matter-energy processes (Ulanowicz, 1997, 2009b).
6. Relations *need not be* conserved—relations *can be* created and destroyed.
7. During and after Life arises, we imagine existence of the first environment−environment relation plus two new relations: Life−environment relation and the Life−Life relation. This is consistent with laws of thermodynamics in that the "something new" created is a set of two new relations.
8. The two new relations were created inside the former undifferentiated system (environment).

These propositions feed into the synthesis hypothesis of the original, fundamental, and relational nature of Life:

A key quality of the Life−environment relation is its integration with events, processes, trends, dynamics, and further articulated relations that all (on average, in the sum, or when taken as a whole) serve to support, sustain, aid, and increase the quantity, quality, and complexity of Life. That is, the Life—environment relation is inherently "good" for Life and is beneficial to the continued existence and operation of Life in the environment. We see this to be true from the origin Life and for nearly 4 billion years afterward.

Here, we again invoke *values* where we hypothesize that a defining characteristic of Life—which again cannot be fractionated from, and must be considered in terms of, the Life—environment system as a whole—is to exist and operate in such way to be supported and aided by, to create and maintain a "good" (or positive) relationship with, the nonliving environment.

It is critical to be clear that this value basis centers on the continued existence or survival of *Life*—life itself, life as a unified whole, including "sustained life"—and not an emphasis on the continued existence or survival of an individual organism, species, or other subset of Life (as in to consider strictly "discrete life"). The same applies to the idea of the environmental function or role as being "good" for Life. This value judgment, too, would only apply to Life as a unified whole, which is able to continue and improve in concert with continued improvement in soils, atmosphere, and environmental Life-support capacity, and not to individuals and organisms who inevitably suffer death, nor to species, families, genera, communities, and ecosystems which inevitably suffer extinction or annihilation.

We admit some difficulty with this approach to analyze and study the idea that the environment can act or be "good" toward Life, and even more challenging to address how Life can be "good" for the environment. The main difficulty arises because it is not obvious how to define a winning, "good," or positive impact on the environment, at least not in terms of an isolated relation. Qualities we might consider good are ones that increase both the energetic efficiencies and total energy use as well as pay-it-forward to offer other parts of the system some share of these benefits. These include relations that can "squeeze" more value and usefulness out of a particular energy unit, such as aerobic respiration getting closer to ground state than anaerobic respiration, lifting the system along a higher gradient. These qualities can often be observed in the overall system complexity and diversity. The "goodness" does not emerge until the relation is processed, or in some way vetted, by the system as a whole. There is almost a tautological aspect to this in that relations that support the good are those that are favored and selected for and those that do not are left behind and forgotten. Conventional wisdom seems to hold that "environment" has no basis for value, good/bad, or right/ wrong *on its own*. We see the value-neutral environment in science, and it also appears in philosophy, Buddhism and Taoism, as in this quote from Watts (1959):

> Looking out into [the universe] at night, we make no comparisons between right and wrong stars, nor between well and badly arranged constellations. (p. 5)

This quote refers to extraterrestrial aspects of environment, but many use similar logic in reference to species extinctions, a lack of direction or progress in evolution, climate change, soil degradation and other natural phenomena, and side effects of human actions. All of these aspects of environment are treated as essentially the same as storms, floods, and natural disasters—like either "random events" or "acts of God" which cannot be assigned any clear or solid determination as "good" or "bad."

Thus, we assert the only solidly justifiable way to conceive of win, benefit, and what is good for environment is *self-referential with respect to Life and human life*. By this rationale:

> What is good for the environment is anything that allows the environment to aid Life and by extension humans.

We include reference to human life here, not just out of self-interest, hope for human survival and sustainability, or anthropocentric values. We also see humans as a unique and crucial member species of Life as a whole, with unique capacity and important roles to play in service to Life, as we discuss later in Chapter 9.

Some self-referential definitions have been taboo in mathematics, such as self-reference in sets which Whitehead and Russell attempted to ban in their Principia Mathematica. Self-reference can be pragmatically pathological in math, computing, and some science realms; for example, self-reference in a formula in an Excel spreadsheet, or tautological statements that are trivial. Nonetheless, here we propose that a self-referential definition for value and what is good for environment is fully necessary. Similar self-referential logic is being employed effectively in allied studies where complex systems are treated in non-fractionable ways; we discuss one of these with reference to hypersets in Chapter 6, Lesson 7. (At the same place in Chapter 6, we also work with more basic ideas such as the definition of impredicative logic.)

This starting place for value that we employ in our relational origin of Life scenario—what is good for the environment is something that allows environment to aid Life—is compatible with the concept we asserted in earlier chapters—that Life is the central basis for all value. Again, we see this basis for value as logical and valid: Life is unique relative to nonliving aspects of the environment, it is certainly important and special, it can be destroyed and thus requires care and stewardship, it is shared by all humans, and it is thus an excellent and fitting basis for unity among sciences and among people and nations. Therefore, we choose this value basis as the foundation for understanding both the original win−win Life−environment relation, and the future and ideal win−win human−environment relation.

This framing of the first principles of Life, environment, and a mutually beneficial relationship between them depicts the original differentiation of life and environment not as a clear-cut separation between system and context, and not as an image of a "self" as isolated or fully autonomous from an environmental "context" as "non-self." Instead, this view of the original, fundamental nature of Life focuses on an organic context dependence, associated with continual collaborative co-creation. This essential interdependence is compatible with "dependent co-arising," a central idea of Buddhism (Macy, 1991).

Another way to say this in summary is (1) environmental context is integral to Life ever and always and (2) environmental context is not merely neutral, not just the provider of an arbitrary or dispassionate "natural selection" of winners and losers—environmental context actively aids, augments, and enhances Life from the origin onward. Were it set up the other way—if Life originated and operated as fundamentally organized to go against the grain of the universe, fighting the local planetary context, with a base relation of win—lose antagonism, or opposing all the natural environmental tendencies—the massive, ubiquitous, continual influence of the environment would plausibly have destroyed Life early on.

We could continue with great amounts of additional discussion of corroborating works, ideas, and scientists. Morowitz (1992), for example, acknowledged that the origin of life could be considered to involve planetary processes including what he called "protoecological cycles." He also mentioned synergy between anabolic and catabolic functions. These and many other compatible views exist, but we must end our focus on the origin of Life here and move on to consider a similar perennial question.

HOLISTIC PERSPECTIVE ON "WHAT IS LIFE?"

Building on these new holistic images of Life, including our holistic hypothetical origin of Life story, we next address the fertile ground of the recurring question, "What is Life?" (Schrödinger, 1944; Murphy and O'Neill, 1995). We also document how problematic it is that Life and environment are arbitrarily split apart, from biology textbooks to dictionary definitions to common everyday understanding of Life. Part of this examination involves ambiguity with definitions and meanings of the terms "life" and "ecosystem." We discuss how this ambiguity is a good thing that hints at the holistic way we see best to answer "What is Life?".

We first examine the current mainstream answer to this question, or the general sense of what life is. The dictionary definition of life is one indication of how the term "life" is used and understood. The Merriam-Webster Collegiate dictionary (merriam-webster.com) defines "life" as "an organismic state characterized by capacity for metabolism, growth, reaction to stimuli, and reproduction."

In line with the dictionary, Morowitz (1992) wrote that "all life is cellular in nature ... A cell is the most elementary unit that can sustain life" (however, see below: he said something different with respect to *sustained life*). It seems fair to say that cells and organisms are generally seen as the fundamental units of life. We also use the term *unit-model* of Life when explicitly acknowledging the close integration of science and modeling (as in Rosen's modeling relation, and Patten's work). Thus, as we form an answer to "What is Life?", one part of the answer for the conventional paradigm (in both science and culture) is that *life is a property of a cell or organism*.

When it is defined or characterized in this conventional cell/organism unit-model, life is described as displaying a list of special properties such as metabolism, growth, reproduction, sensing, and acting on environmental stimuli, being made of cells, and containing information molecules such as DNA. While such Life properties are widely familiar, useful to differentiate life from nonliving entities, systems and processes, and logically associated with organismal life, not all textbooks, dictionaries, biologists, scientists, or people agree on which properties should be included in the list, or on the definitions and descriptions of the essential Life properties. Lahav (1999), for example, reported a long list of diverse and often conflicting definitions and characterizations of Life. The Encyclopedia Britannica online (britannica.com) lists five distinct approaches to defining Life based on physiological, metabolic, biochemical, genetic, and thermodynamic considerations. Still more disagreement is seen from a comparison of the attributes of Life emphasized in Campbell et al. (2008) and Ireland (2010), two widely used college introductory biological science textbooks.

Soon we will see that scientists have had a similar challenge to define "ecosystem." But first, consider this high standard as a potential approach for creating or choosing definitions for Life and for ecosystem. Solomonoff (1997) suggests the importance of *operational definitions* for science:

> An operational definition of anything is a precise sequence of physical operations that enable one to either construct it or identify it with certainty. When one can't make an operational definition of something, this is usually an indication of poor understanding of it. Attempts to operationalize definitions can be guides to discovery. I've found this idea to be an invaluable tool in telling me whether or not I really understand something.

We admit our attempts here do not produce fully operational definitions of Life or ecosystem, in Solomonoff's sense of "certainty" above. But, we agree with him that the exercise itself can be a guide to discovery, is valuable for understanding, is important for developing holistic Life science, and helps with the project of realizing human—environmental sustainability.

As we attempt to articulate an operational definition for Life—a means to either construct it or identify it with greater clarity and consensus, if not total certainty—we shine a light on those areas that are most difficult on which to reach agreement. And by extension, the same crux issues that make definition of life (or Life, as we are framing it holistically) problematic hold potential for helping to flush out the root causes of our problems that now jeopardize Life, and the root causes to enable solutions and better protection, sustaining, and stewardship of Life.

While working with Solomonoff's very high standard, and seeking operational definitions of life and Life, we consider Solomonoff's concept broadly. Where he spoke of the need for "a precise sequence of physical operations," we consider the possibility that "physical" can encompass relational as well as material entities, and that mathematical structures or operations may be essential as well. We also leave open the possibility that no "operational definition" can ever yield a perfect, fully precise, or fully complete definition—such an achievement may be impossible given the ambiguity of language and inherent incompleteness of any formal system shown by Gödel (Raatikainen, 2018).

Robert Rosen studied the question, "What is Life?" over many years and via many approaches. He spoke in poetic terms of this great question (Rosen, 1991) and how essentially all of his published scientific work was:

> ... driven by a need to understand what it is about organisms that confers upon them their magical characteristics, what it is that sets life apart from all other material phenomena in the universe. That is indeed the question of questions: What is life? What is it that enables living things, apparently so moist, fragile and evanescent, to persist while towering mountains dissolve into dust, and the very continents and oceans dance into oblivion and back? To frame this question requires an almost infinite audacity; to strive to answer it compels an equal humility. (p. 11)

One of his many contributions was a holistic model of life he called the metabolism-repair model or (M, R)-system (e.g., Rosen, 1958, 1991). He developed this model, focused on the organism scale, via general systems theory, and mathematics using algebra and especially category theory. He associated "metabolism" with basic anabolic and catabolic processes in living cells. He integrated a generalized "repair" function, observing that all metabolic components— enzymes, for example—last and function for a finite life-time after which they cease to function and must be replaced or repaired (Rosen, 1958). In perhaps his most well-known book, *Life Itself* (Rosen, 1991), he used this model to address the question: "What is life?" In this book, he asserted that a more relevant and tractable question is why is an organism different than a machine? He answered these linked questions by showing how life (again, he focused on organismal life), is "closed to efficient cause" unlike a machine (Rosen, 1991). We take the essence of his result to mean that by their special internal relationships of metabolism and repair organisms are self-making or self-causing. See more discussion of Rosen's work on complexity and his relational model of life in Chapter 5—part 3, Principle 5. And, as mentioned below, we also borrow ideas of Rosen (1985) on the fundamental importance of anticipatory behavior for understanding Life, and by our extension here, for building holistic Life science.

We look next at the typical answer to the question, "What is an ecosystem?". As we shift to this angle, we acknowledge we are dealing with different time, space, and organization scales between organisms and ecosystems, and that this issue of scale must be treated directly at some point. The term ecosystem was coined in 1935 in a paper in the journal *Ecology* by the British botanist Sir Arthur Tansley. Tansley (1935) was writing as a response to what he felt an abuse of terms dealing with community interactions that were labeled as being "organismal." His compromise was to make a new word that captured the concept of communities of organisms functioning together within their environment. He defined an ecosystem as "the whole system (in the sense of physics), including not only the organism-complex, but also the whole complex of physical factors forming what we call the environment of the biome …" (Tansley, 1935, p. 299). Note, this definition has held largely intact during the ensuing 80 + years. Tansley (1935) also wrote, "Though the organisms may claim our primary interest, when we are trying to think fundamentally we cannot separate them from their special environment, with which they form one physical system."

The same dictionary cited above (merriam-webster.com) defines ecosystem as "the complex of a community of organisms and its environment functioning as an ecological unit." Thus, ecosystems are understood as interacting and higher order collections of the fundamental units of Life, namely organisms. Putting the two definitions together, we could describe the conventional view this way: life is a property of organisms, and ecosystems are a collection of organisms interacting with each other and with their environment. An extension of this reasoning is that we must keep in mind that the environment is not strictly abiotic but has emerged with Life as expressed in the concept of *conbiota* introduced above. Fath (2015) used conbiota to address the fuzzy boundaries and, in this way, helped to bring into focus the Life–environment relation that is our focus:

> Therefore, it is at the ecosystem scale that possesses all necessary aspects to sustain life obligatory (Keller and Botkin, 2008). In fact, the life–environment interactions permeate so fully that on a living planet, the very notion of abiota loses its meaning. Life conditions the environmental factors that we typically associate with abiota such as temperature (both local and global), humidity, soil moisture and percolation rates, stream flow, ocean salinity, nutrients concentration, etc. A more apt term would be *conbiota* - the 'physical' environment only makes sense as expressed *with life*. This gives an important clue into the features of sustainable systems. (p. 14)

But, there remains ambiguity over implementation of the concept. The more we look, the more complex the definitions become and the more the distinctions seem blurred. Even merriam-webster.com refers to "functioning as an ecological unit," by which one could interpret "ecosystem" as a bona fide entity that is not divisible, since it is unitary. O'Neill et al. (1987) also spoke to the fuzzy boundaries between Life and ecosystem definitions. In their book, they

> … define ecosystems as the smallest units that can sustain life in isolation from all but atmospheric surroundings. However, one is still left with the problem of specifying the area that should be included.

This is very close to defining the ecosystem as the fundamental unit of Life, as "smallest units" can be considered the simplest or most basic units. Morowitz (1992), despite titling his book "The Origins of Cellular Life," also blended the two concepts, stating that "sustained life is a property of an ecological system rather than a single organism or species." Similarly, Keller and Botkin (2008) wrote:

> To understand how life persists on Earth, we have to understand ecosystems. We tend to think about life in terms of individuals, because it is individuals

that are alive. But sustaining life on Earth requires more than individuals or even single populations or species ... Living things require 24 chemical elements, and these must cycle from the environment into organisms and back to the environment. Life also requires a flow of energy ... Although alive, an individual cannot by itself maintain all the necessary chemical cycling or energy flow. Those processes are maintained by a group of individuals of various species and their non-living environment ... Sustained life on Earth, then, is a characteristic of ecosystems, not of individual organisms or populations. (p. 66)

These ideas fed into our development of the coupled complementary definitions of "discrete life" and "sustained life." (See Chapter 1 and also Lesson 1 in Chapter 6 for full explanation of our discrete/sustained life concept.)

Yet another possible path is to move to the framework of evolution, arguably the strongest basis for integration of biology and ecology. For example, Eigen (1995) tackled the definition of life by defining *a unit of selection*. He listed self-reproduction, mutation, and metabolism as properties of systems that are "predestined" to selection (i.e., possessing the capacity for selection inherently). While these properties more often are associated with cells than ecosystems, in light of Ulanowicz's revised view of evolution via autocatalytic loops, Eigen's approach opens another question. Could an ecosystem (or biosphere)—if considered a holistic organization of physical and chemical dynamics and a unified Life—environment relation—provide an equally or even more robust unit of selection than a cellular, organismic, metabolic, or genetic organization?

To address this, and to develop our answer to the question of what Life is, we employ the same distinction to Solomonoff's "construction" process for operational definitions as we did to Life. We note that one-time construction is distinct from continued construction, and thus, we can speak of "discrete construction" of Life and "sustained construction" of Life as both necessary, as fully interdependent, but also categorically different. It is also useful to consider time scales that discrete construction in the sense of the origination of Life can be seen as a geologic endowment, providing a baseline contribution of bedrock, nutrients, and climate. Overlaying that foundation are processes at the ecological time scale that contribute to sustained construction. Shifting to the present day, we could say that our current global, systemic ecological crisis, and growing threats to our ability to sustain Life, has resulted in greater emphasis on defining sustained construction of Life relative to defining life in a discrete, one-time construction or snapshot approach. Extensions of this discussion will have impact on topics such as artificial life and artificial intelligence.

An interesting result of this approach, linked to our major thesis to integrate Life and environment, is that we end up with a theory and a system in which Life selects its environment (or alters, improves, and causes the environment to adapt) just as much as the environment selects for the most fit Life forms. Yet again, we see that "what is old is new," and that we are able to build on work contributed long ago. Henderson (1913) wrote of the "fitness of the environment":

> The fitness of the environment is one part of a reciprocal relationship of which the fitness of the organism is the other. This relationship is completely and perfectly reciprocal; the one fitness is not less important than the other, nor less invariably a constituent of a particular case of biological fitness; it is not less frequently evident in the characteristics of water, carbonic acid and the compounds of carbon, hydrogen and oxygen than is fitness from adaptation in the characteristics of the organism. (p. 113)

Despite some of his language and ideas relating to mechanism and mechanistic science with which we may now disagree, we credit Henderson with early awareness and appreciation of this holistic Life−environment idea and approach to science which we now employ and promote.

Another essential Life−environment relation we must understand is Life's ability to build gradients in energy and material components, to maintain the gradients so they persist, grow, and increase in complexity, and to harness the gradients for useful work to aid Life. By gradients, we refer to nonrandom structures and varying amounts, stocks, or concentrations of energy as well as key material constituents (e.g., carbon, nitrogen, water, etc.). This gradient building capacity is closely linked to the ways we described above that Life improves the environment over time, as in the atmosphere and soils. Two issues at play are (1) the build-up of gradients and (2) the slow release, use, or dissipation of the gradients. The build-up is a result of autocatalysis, which we covered extensively above. We described Life's inherent growth capacity made possible by connected loops of processes, and positive self-enhancing feedback, causing growth and improvement in all nodes in a cyclic network, including environmental components such as soils, atmosphere, and more. The slow release is achieved through the close coupling of multiple Life processes. Rather than expend accumulated stores of energy or matter, to "blow the load all at once," Life processes exhibit a hierarchical stair-step of smaller and smaller cycles each being driven by the dissipation of the initial energy gradient. Fath (2017) used this simple figure to contrast the quick versus slow release of energy or dissipation of useful gradient (Fig. 4.1).

In summary, our multipart answer to "What is Life?" is based on the more holistic, more synthesis-oriented science we have been developing and promoting. Our answer is still complex, ambiguous, incomplete, and requiring a dynamic collaborative learning process and constant renewal going forward, but it contains these major component concepts we see as useful:

1. Life is not strictly reducible to a cell or organism nor to a process or property of a cell or organism.
2. Life is not strictly reducible to an ecosystem nor to a process or property of an ecosystem.

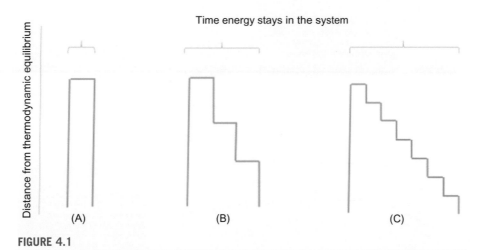

FIGURE 4.1

Visualization of an energy pulse that is degraded at different rates depending on the number of intermediate reactions that are coupled to the gradient utilization: (A) no coupling, (B) moderate coupling, and (C) extensive coupling.

Redrawn from Fath, B.D., 2017. Systems ecology, energy networks, and a path to sustainability. Prigogine Lecture. Int. J. Ecodyn. 12 (1), 1–15.

3. Life is complex and thus irreducible; there is no "single largest model" in the sense of Rosen (1991) and no single best, complete, sufficient, unambiguous unit-model by which to frame and study Life.
4. Definition, construction, and understanding of Life requires at least three unit-models—cell/organism, community/ecosystem, and biosphere.
5. Operating in concert, these three integrated and interdependent Life systems are able to construct (create) Life and a Life-supporting environment indefinitely over time.

We see these three unit-models as *holons* (Koestler, 1968)—wholes which are integral parts of larger wholes. To unify these three unit-model holons, we go more in-depth into a proposed integrated multi-model using hypersets, in Chapter 6, Lesson 7. For now, we close this section with additional quotes, ideas, and corroborating work that help to paint a picture of this new holistic, irreducibly complex, self-referential, self-enhancing Life—environment system we seek to envision and depict, so that we may learn from it and mimic its success. We also share another diagram and an artistic-and-scientific representation of the organic nestedness of these multi-scale Life systems.

An amazing feature we continue to see is how Life emerges and self-sustains through a type of self-organization such that each participant "thing" is simultaneously "doing its own thing" and "doing its own thing to fit together." Fig. 4.2 shows a small subset of this cooperative organization between a generalized plant or plant-like function and generalized animal or decomposer function using network

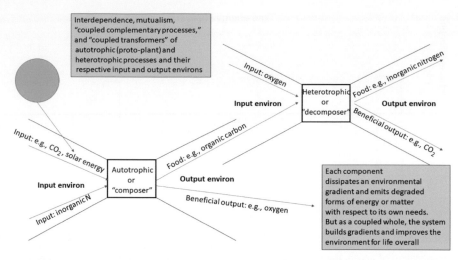

Interdependence, mutualism, "coupled complementary processes," and "coupled transformers" of autotrophic (proto-plant) and heterotrophic processes and their respective input and output environs

Input: oxygen

Heterotrophic or 'decomposer'

Food: e.g., inorganic nitrogen

Input environ

Output environ

Beneficial output: e.g., CO_2

Input: e.g., CO_2, solar energy

Food: e.g., organic carbon

Autotrophic or "composer"

Output environ

Input environ

Input: inorganic N

Beneficial output: e.g., oxygen

Each component dissipates an environmental gradient and emits degraded forms of energy or matter with respect to its own needs. But as a coupled whole, the system builds gradients and improves the environment for life overall

FIGURE 4.2

Environs doing their own thing and doing their own thing to fit together.

environs of Patten. In this figure, outputs and byproducts of one environ (which is a multi-scale model and can apply to an organism, species, ecosystem, or other Life unit) serves both as an output and as input to another network environ. Since input and output environs of unique entities can overlap, the output of one structures the input of another. This mutualistic interplay aids Life's ability to build gradients, not just dissipate them, when many linked environs connect in loops as in the autocatalytic loops of Ulanowicz. Either individual component may be seen to dissipate a gradient, or degrade higher quality input energy or material. However, as a combined and complex functional unit, and due to autocatalysis of the loop system as a whole, energy and material gradients—perhaps along some third dimension not associated with either interacting component—can be built, slowly used, and leveraged to aid key Life functions. Vertical structures that Life spontaneously constructs in soils, and differential concentrations of gases Life spontaneously constructs in vertical layers in the atmosphere, are our best examples.

Kauffman (2011) has written of this characteristic, holonic, unity stating: "The function of each task is its role in the reproduction of this Kantian whole." He also wrote:

> ... biological evolution concerns Kantian wholes, where the whole exists for and by means of the parts and the parts exists for and by means of the whole.

And

> A collectively autocatalytic set of peptides, as exemplified by Gonen Ashkenazi of Ben Gurion University and his nine peptide collectively autocatalytic set, is a clean example of a Kantian whole, achieving a closure in "catalytic task space," where all reactions requiring catalysis are catalyzed by members of the nine peptide set.

We will return to Kauffman's "Kantian whole" and seek to clarify the ideas and sharpen the image, which we see as perhaps *the* crux idea of Life in a nutshell. As we unpack this complex concept and make sense of it, we can proceed to make a new holistic Life science, with testable hypotheses and predictions, and ultimately with application to successful human sustainability.

Another compatible and supporting quote, "The results of systemic operations are at once more systemic operations," comes from Moeller (2006) in a book about Niklas Luhmann. This quote, and both authors, helps reinforce the essential interplay between system independence and interdependence.

In a collaborative contribution to aid prior work, Sarah McManus created the image below (Fig. 4.3A) depicting the three unit-models we have described as necessary holons—Life simultaneously as organism, ecosystem, and environment (or biosphere). While this image is two-dimensional, McManus also created an image (Fig. 4.3B) and instructions for folding this image into a "hexaflexagon" that becomes a three-dimensional structure that can then be folded repetitively as a spinning cycle. This was her brilliant translation of our idea of using hypersets as integrated multi-model of Life (Fig. 4.4).

McManus's art work, images, and three-dimensional form provide an excellent link to the graphical work of M.C. Escher, whose "Whirlpools," "Drawing hands," "Ascending and descending," and similar works help to invoke the sense of paradox, folding, and complex interplay of dimensions that we consider necessary conceptual tools and terrain for understanding Life in full holistic complexity. These artistic expressions also link to increasing scientific evidence and observations of fuzzy boundaries and complex relations, such as growing awareness that ecosystems are enfolded within organisms, such as in the gut microbiome. These cases of multi-scale Life are also known to be of practical

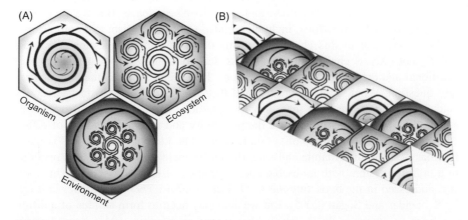

FIGURE 4.3

(A) Three unit-models of Life by McManus (2013). (B) A pattern that can be folded to form a dynamic, cyclically foldable hexaflexagon. For explanation of the hyperset equation, see Chapter 6.

Printed with permission of the author.

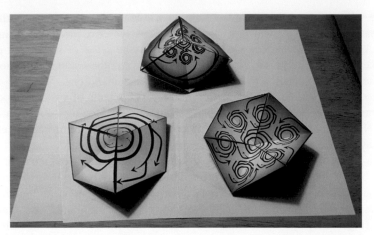

FIGURE 4.4

Three different folded configurations of the single hexaflexagon of McManus showing from left to right the organism, environment (or biosphere), and ecosystem holons.

Printed with permission of the author.

importance, such as relations between the ecological dynamics of gut microbes and human health outcomes.

We could go on and broaden the exploration even more; we could examine fractals with their self-similar structure at many scales; we could explore Rosen's (1991) metabolism-repair model for the complex life of an organism; we could add to Ulanowicz's idea of autocatalytic centripetality (pulling inward) by showing entwined centrifugality (radiating outward) in Life systems, invoking ideas of waves or fields that interpenetrate and mesh in phase space. But we save these explorations for future efforts, including related topics on holism and complexity.

We have seen in this chapter how this deeply profound question, "What is Life?", not only is challenging to address, let alone answer, but it also invokes additional and equally challenging questions. Patten has explored the complementary question, What is environment?, and this too has led to important advances. And, in this book, we have seen how the question, What is death?, also must be addressed. We have also discussed the possibility and potential benefits of differentiating between two forms of systemic death. The idea of "heat death" is likely familiar as a thermodynamic end state driven by increasing entropy and resulting in a uniform system with no usable energy gradient, like a diffuse gas. This topic was addressed in the book *Into the Cool: Energy Flow, Thermodynamics and Life* by Schneider and Sagan (2005). But we also may need to form an idea of a different form of systemic death driven by *syntropy* (Fantappiè, 1942; Szent-Gyorgi, 1977; Fuller, 1979), characterized by increasing orderliness, yet also a bland configuration in its uniformity, until an endpoint like a frozen crystalline lattice incapable of any dynamics or Life process. Syntropy is another term for negentropy, but Fantappiè and others developed concepts associated with syntropy beyond those of negentropy. Life may then best be seen as a perpetual balancing act between these two attractor basins of heat death at one extreme (too little order or constraint) and some other kind of death at the other extreme (too much order or constraint). With analogies to matter, this makes Life more akin to a liquid phase like water, and with process and potential that is fluid, dynamic, able to change, reconfigure, evolve, and grow. See Chapter 6, (Lesson 6) for examination of Ulanowicz's related ideas and quantification of the balancing act between order and flexibility as a universal pattern in ecosystem networks.

A last linked idea for now is the implication that a holistic idea and image of Life will also necessarily alter our concept and model of the human self. This relates to the concept of the "ecological self" of Næss (1989). Just as we have shown that the choice of drawing a boundary, consciously choosing what is inside versus outside any system of study or any concept we can frame, is critical, our sense of self also depends on the boundary of the system being modeled or considered. This is not purely a metaphysical topic, such as a rhetorical question like "Who am I?". Instead, a newly formed sense of the human self holds potential as an alternative core strategy for linking humans, our values, our behaviors, and the ultimate outcomes of whether we will be able to achieve social and environmental sustainability. A healthy sense of self—where by healthy we refer to the capacity for social and environmental sustainability and the care of Life-support systems and services—would plausibly entail an individual (as well as a community, nation, or society) that recognizes "self" in both discrete form (bounded by one's skin, or by one's species) and in extensive form (with wider boundaries integrated with Life-support systems and critical environmental context). This dialectical and irreducibly complex, even ambiguous sense of self thus would enable the individual, community, society, or species to remain healthy in whole, in holistic individual + context fashion. This holistic and multi-scale sense of self, linked to a holistic science of Life, is an aspect of the human wisdom and maturity to which we seek to contribute.

Our next step is to translate the many holistic concepts and supporting work from previous scientists into a set of core founding principles for holistic Life science.

Reforming reductionism with six core principles

5

CORE PRINCIPLES OF HOLISTIC SCIENCE AND LIFE—ENVIRONMENT SCIENCE

We interpret the evidence and concepts in Chapter 3 and Chapter 4, to mean that living systems are inherently skillful at operating in ways that preserve and even enhance the environmental context in which they exist. The fact that our industrial era science, technology, policy, and shared cultural ideas have not yet managed to recognize, understand, mimic, reproduce, and achieve this essential self-enhancing Life—environment relation thus informs our need for science reform.

We next describe core principles of a new holistic science to balance reductionism, repair fragmentation, and unify fields splintered by hyper-specialization. We have set the basis of value on Life for both society and science; we have strategies for choosing system boundaries; and we have explored the conceptual challenges of defining Life and understanding the origin of Life. We next show the need for, and propose new foundations for, science in service to Life. We propose six key characteristics and qualities that summarize that this science:

1. Is consciously, intentionally, and transparently value-based centered on the value of Life;
2. Is anticipatory and accelerates the pace and process of scientific change, including paradigm shifts, toward the ultimate goal of a sustainable human—environment relation and Life—environment relation;
3. Balances and synergizes holism with reductionism and synthesis with analysis;
4. Equally emphasizes internalist and self-referential as well as objectivist perspectives;
5. Is complex (in the sense of Rosen, 1977, 1991, 2000) and is able to reconcile seeming opposites and handle multiple scales and fluid boundaries of focal entities;
6. Is radically empirical with constant capacity for questioning, challenging, and transforming ingrained assumptions and structures, especially whenever these distract scientific attention and resources from those topics of greatest benefit to humanity and Life.

Foundations for Sustainability. DOI: https://doi.org/10.1016/B978-0-12-811460-5.00005-4

Taken as a whole, we see this set of guiding principles to be internally consistent, and with all of the principles supporting the others. We predict that the impact of future development and adoption of such principles would be unifying instead of dividing, would increase stabilizing trends instead of fragmentation, and would enable large-scale focusing of scientific enterprise on those topics of most importance to humanity and to Life. These guiding principles would help us steer a steady course toward solving our current global ecological crisis and preventing against wild wandering paths chasing after whims and current hot topics. This is a proposal for an evolutionary leap for science itself—we envision a science that grows beyond its original meaning of knowledge, and beyond current practices to inform government and decision makers. Holistic Life science takes on responsibility for success and acts in service to Life from a wisdom and maturity integrated with the traditional humility and skepticism that have been science's strengths. We now go into more detail for each of these six principles.

Principle 1. Holistic Life science has a value basis centered on the value of Life.

This principle is the focus of Chapter 2. The only idea we mention here is that the holistic science we propose can be considered "value neutral" given that the value of Life is universal among all people, applies to everyone equally, and introduces no value bias of any detrimental kind in the sense of any "conflict of interest" or invocation of identity politics. To say this another way, the only value basis that is discriminated against or given less priority is a value system centered on Death (systemic Death, the death of all living things), which by our definition has no advocates, adherents, proponents, or anyone with any "standing" in the legal sense to be discriminated against, marginalized, or left out. We can easily say if anyone has an objection to "speak now or forever hold your peace" and be certain that no one will speak. Thus, we also see this value foundation to offer no weakness or vulnerability to any attack to the strength or rigor of the science involved; the value we are imbuing is essentially value neutral, as it is value generic, or value universal.

Principle 2. Holistic Life science is anticipatory and oriented toward a sustainable human—environment relation.

This project to reform science and its foundations is inherently anticipatory. At each step, we are thinking and looking ahead with a clear idea in mind of the ultimate outcome and "system of solutions" working together in the desired future. Anticipatory science is different from predictive science—it does not merely foresee but acts in advance. This means we also have in mind that the innovations in science philosophy and theory we develop in this chapter must have real capacity to translate fully into innovations in science applications and technology, aspects of which we address in Chapter 8. And, these applications and technologies must also feed forward effectively to achieve the outcomes we imagine as our "problems solved" and "relations healed" scenario to end the current human—environment crisis, without, or at least with minimal, unintended, or unanticipated negative side effects from these solutions. Positive side effects and unintended consequences—similar to those net impacts by which Life has improved its environment over time—are possible as well as consistent with value and respect for Life.

Following Ulanowicz and others, we can look forward and look deeply to assess our current situation, to see the need for a paradigm shift. By following Ulanowicz, we mean not only to pursue a paradigm shift but also to do so with humility and deliberation. In the Preface to his 2009 book, *A Third Window*, Ulanowicz recounts how he began to realize that his work, and the science principles grounded in an ecological perspective he was helping develop, is indeed a new paradigm. After a colleague suggested to a graduate seminar class on the philosophy of science that Bob's presentation of his work represents a new paradigm, Bob wrote of his reaction and thoughts:

> Initially I was irritated, given my aversion to overuse of Kuhn's word paradigm. There followed, however, a tinge of excitement at the possibility that maybe I had not fully appreciated how much the ecological perspective can alter how we see the rest of the world. Perhaps ecosystems science truly offers a new angle on nature (Jørgensen et al., 2007). Hadn't Arne Næss (1988) proposed that "deep ecology" affects one's life and perception of the natural world in a profound and ineffable way? Although I am not averse to the transcendental, I do nevertheless expect scientists to exhaust every rational approach to phenomena before abandoning them as ineffable. (page xix)

As we hope our evidence and citations of corroborating scientific research show, we also work to "exhaust every rational approach" in our efforts to understand the world. We are also compelled by the excitement and possibility Ulanowicz mentions, and, while not only or primarily transcendental, we do see the potential revolutionary change from an antagonistic to a synergistic human—environment relation as wonderful, hopeful, and inspiring.

In addition to looking to future promise to guide science reform, we can also learn from the past and seek to turn this learning into a proactive and anticipatory science and technology. In Western Maryland, Appalachia, and other regions referred to as "coal country," we continue to live with and pay heavy costs of the legacy of energy development based on coal mining. We still suffer the side effects and unintended consequences of mining done over 100 years ago—the land subsides under our buildings as old wooden posts rot and underground mines collapse, and our water runs bright orange due to acid mine drainage and the iron and other minerals it leaches out of the disturbed layers of rocks and minerals. Energy development need not be done in this way, and a forward-thinking science can assist with forward-thinking technology and applications. We return as always to see the highest role model in Life—the ideal side effects and unintended consequences can lead to benefits like oxygen atmosphere, soils, and the ozone layer—if we have organized ourselves in proper relation to the planetary environment.

The precautionary principle is important to mention here and to adopt going forward. To utilize the precautionary principle in all deliberations would be compatible with valuing Life and Life-support ever and always as a top priority. As part of any plan, design, or action, a holistic approach to any new work would scan forward to test scenarios and search for risks to Life and Life-support. If any such threats to Life appear in this scan of future impacts, then one would necessarily redesign or alter the plan to avoid that risk. An ability and habit to be "risk averse" with respect to Life value would drive the abilities, habits, and skills to identify risks and act in advance to prevent them from coming to be.

As we see in many damaged ecosystems and environments, prevention is preferable, less expensive and causes less human suffering than cure. Beginning around 2000, in the process of Chesapeake Bay restoration, advocacy, science, governance and in the press and community, estimates for full restoration of Bay health started at $19 billion and then were revised a few years later to $30 billion, and then people stopped making and publishing such estimates. The years of this receding goal have been dominated by the prior reductionist, objectivist, mechanistic science paradigm including central ideas, theories, and definitions within biology and ecology in which Life has been separated from environment. But, this is the wrong model to address the problem and assess the cost. It is not as though someone, say Warren Buffett or Bill Gates, could just airdrop $30 billion and the problem is fixed. The monies spent will be to revitalize self-healing autocatalytic cycles that restore and regenerate ecosystems and communities dependent on the healthy functioning of the Bay. Once invested, this will reap dividends in years to come, but not in a Wall Street financial sense, but in that the place will become home to and supportive of sustained Life for humans and nature. Only by refusing to accept defeat and failure, by insisting on successful achievement of Chesapeake Bay (and all local ecosystems) restoration and human–environment sustainability, can we muster the will and energy needed to transform our science, values, and society.

The principles we propose and the emphatic focus on success with sustainability may be considered as most applicable to a transitional period in near-term human history. We see these principles, the science methods, and other concepts in this book as necessary to achieve a successful transformation of both science and society. Once this transformation is complete, and the crisis averted, and once we are well on-track with a win–win human–environment relation, then these principles and the associated paradigm we propose could be reexamined and reformed toward new primary focal needs. Given that we see Life value and Life science as of perpetual importance, we would advocate and hope that any future waves of science reform or paradigm shift would continue to ensure Life–environment health and quality going forward.

We seek to make these principles mutually supporting; note that this anticipatory principle is consistent with principle 1—Life value basis. And, see below how it meshes closely with principle 4 related to internalism and self-reference. Combining these, we propose that science itself should operate with a "pay as you go" ethic, not leaving a debt or mess for future generations. In fact, properly done, anticipatory science will not only avoid leaving messes but also will leave positive conditions and opportunities.

A conceptual angle on anticipatory systems we employ follows Rosen (1985), whose book, *Anticipatory Systems*, asserted that anticipatory behavior is unique to living systems and Life forms. He used anticipatory in the sense of a system in which the future determines or influences present action. By aligning our proposed holistic Life science with this meaning of anticipatory, we seek to mimic Life as we build this science of Life.

These proposed principles borrow much inspiration and insight from Stephen Covey's (1989) book, *Seven Habits of Highly Effective People*, and his subsequent works on principle-centered leadership. Covey's body of work sought to anchor human behavior on a foundation of "natural law" very much akin to our proposal to mimic nature and ecological systems. He also promoted an ethic of leading by example and wrote of the unity of right thinking and right action. Covey's (1989) first two habits are "be proactive" and "start with the end in mind." Our second principle of anticipatory science mirrors and seeks to adopt these two habits. Covey's fourth habit is "think win/win" and his sixth "synergize." We seek to follow these habits in the remaining four principles.

Principle 3. Holistic Life science balances and synergizes holism with reductionism, and synthesis with analysis.

Mainstream science is primarily reductionist with emphasis on analysis. Many trends and workers are helping to remedy this overemphasis and to increase the capacity of science for synthesis. We support and join these allies and seek to build on work of holistic scientists like Bateson, Ulanowicz, Patten, Lotka, Bohm, Goerner, and many more. We also borrow relevant principles from Eastern philosophical schools and traditions such as Buddhism and Taoism, ancient sources of wisdom with potential to make Western science more holistic and thus help us to see, understand, and provide stewardship for Life–environment as a unified whole. As recommended by Billings (1952), we propose here that in holistic Life science "... analysis must be followed by a synthesis of the total results." In fact, the holistic and synthesis approach can also come first, followed later by analysis as needed.

Gregory Bateson contributed holistic work and ideas touching on our focal topics of Life, environment, thinking, and sustainability. In a talk in 1970, the year of the first Earth Day and creation of the US Environmental Protection Agency, he said (Grossinger, 1978):

> We face a world which is threatened not only with disorganization of many kinds, but also with the destruction of its environment, and we, today, are still unable to think clearly about the relations between an organism and its environment. What sort of thing is this, which we call 'organism plus environment?' (p. 30)

This simple yet profound observation calls attention to the need to understand better not only both organism (what is Life?) and environment (what is environment?) but also the ever present, inseparable, and obligate relation that exists between the two. In our prior paper (Fiscus et al., 2012), and in this book, we seek to help us all "think clearly" to get at the same basic ideas—what is the system that unifies Life and environment, and can this approach help explain and reverse our destruction of the environment? Bateson went on:

> It is now empirically clear that Darwinian evolutionary theory contained a very great error in its identification of the unit of survival under natural selection. The unit which was believed to be crucial and around which the theory was set up was either the breeding individual or the family line of the subspecies or some similar homogeneous set of conspecifics. Now I suggest that the last hundred years have demonstrated empirically that if an organism or aggregate of organisms sets to work with a focus on its own survival and thinks that that is the way to select its adaptive moves, its "progress" ends up with a destroyed environment. If the organism ends up destroying its environment, it has in fact destroyed itself. And we may very easily see this process carried to its ultimate *reductio ad absurdum* in the next twenty years. The unit of survival is not the breeding organism, or the family line, or the society. (p. 32)

For evidence of the penetration of this overly reductionistic error in thinking about evolution, look no further than the images used to present evolution in introductory biology textbooks. A typical image shows the branching pattern from only the organisms' evolutionary perspective absent of the environmental context or relations in which the coevolutionary processes occur. Returning to Bateson, he continued:

> The flexible environment must also be included along with the flexible organism because, as I have already said, the organism which destroys its environment destroys itself.

> The unit of survival is a flexible organism-in-its-environment. (p. 32)

Bateson, in this same talk, switched from a focus on "life" to focus on "mind" and again sought holistic integration (Grossinger, 1978). Just as analysis can be a stepwise process of splitting and studying subsets in isolation, synthesis can be a stepwise process of ever fuller integration. By Bateson's approach to seek and employ the "pattern that connects" (a phrase he wrote later, Bateson, 1988) and by generalizing the idea and process of a trial-and-error system, Bateson first conceptually unified life and environment into a unit of survival or evolution and later proposed further unification and wholeness.

The talk quoted above by Bateson was titled "Form, Substance and Difference" and was the 19th Annual Alfred Korzybski Memorial Lecture, January 9, 1970, at the Oceanic Institute, Hawaii. From the book Ecology and Consciousness (Grossinger, 1978). From Bateson, who worked in anthropology, cybernetics, interdisciplinary conceptual studies, and other science fields, we turn to holistic ideas from Eastern spiritual and philosophical traditions.

Alan Watts played a role in bringing many of the views of Eastern religions, Zen Buddhism, Indian, and Chinese philosophies to the West in the 1950s and 1960s. The quotes that follow are from his book, *The Book: On the Taboo Against Knowing Who You Really Are* [1989, Vintage Books (after the 1966 original)]. Watts, though working in a very different field, gave a similar perspective as Bateson's and our own. He wrote of the link between scientific reductionism and human thought process using an analogy of trying to understand a cat walking by while restricted to the view of looking through a narrow slit in a fence:

> The narrow slit in the fence is much like the way in which we look at life by conscious attention, for when we attend to something we ignore everything else. Attention is narrowed perception. It is a way of looking at life bit by bit, using memory to string the bits together — as when examining a dark room with a flashlight having a narrow beam. Perception thus narrowed has the advantage of being sharp and bright, but it has to focus on one area of the world after another, and one feature after another. (p. 31)

And a bit later, with another analogy of brain process like the scanning of radar:

> But a scanning process that observes the world bit by bit soon persuades its user that the world is a great collection of bits, and these he calls separate things or events. We often say that you can only think of one thing at a time. The truth is that in looking at the world bit by bit we convince ourselves that it consists of separate things, and so give ourselves the problem of how these things are connected and how they cause and effect each other. The problem would never have arisen if we had been aware that it was just our way of looking at the world which had chopped it up into separate bits, things, events, causes and effects. We do not see that the world is all of a piece . . . . (p. 32)

Watts also applied holistic thinking and Eastern philosophical perspective to life, humans, and the environment. He wrote:

> . . . technical progress becomes a way of stalling faster and faster because of the basic illusion that man and nature, the organism and the environment, the controller and the controlled are quite different things. We might 'conquer' nature if we could first, or at the same time, conquer our own nature, though we do not see that human nature and 'outside' nature are all of a piece. (p. 51)

And later with ideas that align with attempts at scientific definitions, models, and descriptions of Life:

> The whole is greater than the sum of its parts if only for the fact that a scientific description of the body must take account of the order or pattern in which the particles are arranged and of what they are doing.

> But even this is not enough. We must also ask, 'In what surroundings is it doing it?' If a description of the human body must include the description of what it, and all its 'parts,' are *doing* — that is, of its *behavior* — this behavior will be one thing in the open air but quite another in a vacuum, in a furnace, or under water.

> If, then, a definition of a thing or event must include definition of its environment, we realize that any given thing *goes with* a given environment so intimately and inseparably that it is more difficult to draw a clear boundary between the thing and its surroundings. (p. 67—68)

And finally, indicating that ignoring this wholeness can be deadly:

> We cannot chop off a person's head or remove his heart without killing him. But we can kill him just as effectively by separating him from his proper environment. This implies that the only true atom is the universe — that total system of interdependent 'thing-events' which can be separated from each other only in name. (p. 69)

Similar to Bateson, Watts took his Buddhism-inspired holism to a logical extreme and unified the entire universe. And, similar to Bateson and Ulanowicz, he wrote of how this full wholeness necessitates unity of spiritual and material realms as well. The "true atom" he mentions above he later identifies as the self and the soul. We focus our strategy for paradigm shift and successful achievement of human—environment sustainability on Watts' observation that "we can kill him just as effectively by separating him from his proper environment." We take the "him" in Watts' statement to symbolize modern industrial humankind, and the reductionistic mechanistic science "he" employs. We point to our biology and ecology textbooks and shared ideas and definitions as currently "separating him from his proper environment." If this is true, then reconnecting humans to their proper environment should lead to healing of wounds and a healthier Life going forward. This demands the promotion of an integrated, holistic science that is in need of good teachers and good textbooks.

Watts also reminds us of the value of environmental science, ecology, philosophy of science, sustainability studies, and other holistic, interdisciplinary, and integrative fields. These fields have built methods, habits, and infrastructure by which we can do more than see only isolated bits of the world illuminated by specialized sciences serving as narrow flashlight beams or restricted views from slits in a fence. As we widen the view and take in more, and as we connect the patterns like Bateson, we begin to see important new relations such as the features shared by all the symptoms of environmental degradation in Fig. 1.1.

In diagrams and text from our prior paper (Fiscus et al., 2012), we used a thought experiment of tracing the flow and movement of an atom of carbon as a means to highlight the holistic unity of Life and environment. The flow path traces three major material cycles and our three integral holons (organism, ecosystem, and biosphere) and weaves them together. As this imagined carbon atom

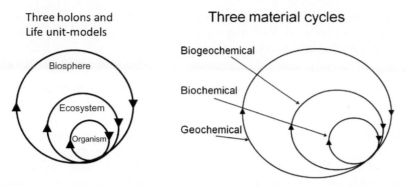

FIGURE 5.1

Three integrated and interdependent holons or unit models of Life, and their associated three materials cycles.

Modified from Fiscus D.A., Fath, B.D., Goerner, S., 2012. The tri-modal nature of life with implications for actualizing human-environmental sustainability. Emergence Complexity Organ.14 (3), 44—88.

moves between organism, ecosystem, and biosphere, it provides corroboration for principle 3 of holism here, and for principle 5 of multi-scale complexity below.

Tracing the flow of any single atom involved in Life process over an extended period of time shows three material cycles to be fully unified. This tracing of elements has been done quantitatively and scientifically by use of radioactively "labeled" atoms. Patten and Witkamp (1967) and Neal et al. (1967) reported work with radioactive tracers to understand structure, function, and materials movement in ecosystems. The general processes are basic enough that we used a thought experiment to convey the main idea, as inspired by Harding (2006), and used the simple system diagram (see Fig. 5.1). In 2012, we wrote:

Imagine an atom of carbon fixed into plant sugar from the atmosphere during photosynthesis. This carbon atom bound into glucose might move to another area inside the same plant (along a portion of the internal biochemical cycle). The glucose might then be metabolized to provide energy for a physiological process in the plant, and the carbon atom might then be respired and emitted back into the atmosphere where it began. But it might also be that the original carbon atom eventually became incorporated into cellulose, and then entered the soil when that plant died. If so, then it could provide food for another living organism such as one of many leaf-eating insects. Then once embodied in the insect the carbon atom might move locally as the insect travels or disperses, or it could be transformed on up the food chain if eaten by some predator (either route part of a biogeochemical cycle). If the carbon atom became incorporated into the organic matter of the soil, then it might only return to the atmosphere much later (perhaps after hundreds or thousands of years), or it could be washed away and travel far downstream or even to the ocean via actions of rainfall, erosion and stream transport (part of a geochemical cycle). From this *or any similar brief survey of the many possible pathways for our single carbon atom, it is clear that all three major material cycles can and must be involved in any and all life processes.*

Fiscus et al. (2012, pp. 67–68)

We find a similar result if we trace the journey of oxygen, hydrogen, nitrogen, phosphorus, sulfur, or other elements essential to Life and provided by the environment. Continual turnover of elements in organisms and ecosystems is well known. The separation of the three major material cycles (biogeochemical, biochemical, and geochemical)—much as the separation of the three associated unit models of Life (ecosystem, organism and biosphere)—is valid "for discussion purposes only" or as a temporary analytical step to be followed by a necessary synthesis step. Close examination of real systems (by imaginary means like thought experiments as well as scientific tracer studies) shows all three cycles to be inextricably interwoven.

Network analysis research along these lines quantified the degree by which ecological networks redistribute matter throughout all compartments—a phenomenon called "resource homogenization" (Borrett and Salas, 2010; Fath and Patten, 1999). Patten (2016a) has also written of "network nonlocality" and "network enfolding" as two holistic scientific concepts needed to understand Life and sustainability (these are two of many holistic concepts derived and quantifiable from ecological networks and systems ecology; see more in Chapter 6). Of network enfolding, Patten wrote ". . . the environments of systems enter those systems as inputs, and these are progressively incorporated into the fabric of the system by network enfolding" as inputs move through and transform the network via interior flows between compartments. He also wrote that the large set of flows along interior pathways, calculated via an infinite power series in matrix math, "achieve network enfolding, a process that more than any other is the source of systemic holism" (Patten, 2016a, p. 72).

While Patten's technical terminology may seem hard to understand (even if challenging, we recommend everyone read his work), it may help to visualize a typical food web such as that depicted in Fig. 5.2. Entitled "food chain" and only showing three players in a subset of a complex food web, this image is powerful and helps understand network enfolding and systemic holism. It portrays how, as food for the fish, a fly becomes incorporated into the very being of the fish, both of whom then become embodied in and help to form the bear. What is not shown overtly, but we can add with imagination (and we have even asked the artist, Jan Heath, to help with this in a revised "food chain" print) is that flesh (matter and energy) of the bear can also be consumed by and thus be incorporated into the being of the fly. This turns this chain into a cycle leading to the large numbers of cycling pathways that Patten and others use infinite power series to calculate. (Note, this is not in defiance of the second law of thermodynamics as the dissipation at each transfer is lawfully accounted for, but the molecules do cycle around and around.) This cycling of matter and energy that repeats many times, linked to

FIGURE 5.2

Artwork of Jan Heath, entitled "food chain." Used with permission of the author.

the very real way we think of as "you are what you eat," illustrates how even our categories of bear, fish, and fly mask an underlying integration via centuries, millennia, and longer cycles of Life process making them all also unified in a larger Life—environment whole.

Shakespeare spoke of this food-linked holism, too, and even applied it to people. Hamlet says (Eliot, 1914):

> We fat all creatures else to fat us, and we fat ourselves for maggots. Your fat king and your lean beggar is but variable service, two dishes, but to one table; that's the end.

And soon after:

> A man may fish with the worm that hath eat of a king, and eat of the fish that hath fed of that worm . . . to show you how a king may go a progress through the guts of a beggar.

In this respect, network enfolding and ecological holism may be seen to serve as an equalizing influence—the worm, king, fish, and beggar all end up on the same level playing field. The material flow is, in a sense, homogenized and shared throughout the Life—environment system.

It is possible to integrate and see larger wholes via science, visualization, and capacities of the human mind or via caring, compassion, extension of rights and thus capacities of the human heart, ethics, and morality. Roderick Nash, in his book, *Rights of Nature* (1989), charted how the concepts of rights, equal standing, and intrinsic value can be seen to have expanded over human history. Rights originated from narrow self-interest, such that early teaching and tenets focused on people—people interactions (which was largely settled, at least in theory, with some version of the Golden Rule, to treat others as you treat yourself). These rights passed later to human ethics that embraced tribes, families, and regions, and to other humans as through ending slavery and establishing civil rights, in which the challenge is dealing with people—group interactions. Democratic and legal institutions have rules in place, but we still struggle today with instances when the rights of an individual conflict with the rights of many. In recent decades, concern has been extending to rights of nature and other species—people—environment relations—as codified in the Endangered Species Act and other shared cultural documents. Leopold's *Land Ethic* goes a long way in guiding our actions in this realm, and an extended version of the Golden Rule avers to leave the world better than you found it. This expansion of rights has paralleled the development of the field of Environmental Ethics and the work of Arne Næss on Deep Ecology mentioned earlier.

These unifying processes are the complement of the perhaps more often used processes of splitting, dividing, and fragmenting—the ways science fragments into hyper-specialized subdisciplines, or the way that people at times listen to fear or ignorance to see themselves as different than and threatened by other humans based on superficial differences such as skin color, language, or lifestyle choices.

Expressing a similar concept, a famed scientist speaking of religion in a letter to a grieving father, Albert Einstein wrote (Calaprice, 2005):

> A human being is a part of the whole, called by us "Universe," a part limited in time and space. He experiences himself, his thoughts and feelings as something separate from the rest—a kind of optical delusion of his consciousness. The striving to free oneself from this delusion is the one issue of true religion. Not to nourish it but to try to overcome it is the way to reach the attainable measure of peace of mind.

There are scores of other workers who have contributed unique as well as shared concepts that support the value of holism and synthesis, and they span many fields. Goerner et al. (1999), in a book similar to this one seeking foundations for change for sustainability, wrote of "a Great Ordering Oneness" that manifests itself in universal patterns, "sacred geometries" shared by both physical and living structures, and web dynamics, a Oneness that as a whole "weaves order into every nook and cranny" of the universe. David Bohm (1995 after 1980) echoed very similar ideas in his work on "wholeness and the implicate order," in which he sought to change physics but also saw the negative impact of fragmentary thinking on human life. Palmer et al. (2010) wrote, "We are being called into a more paradoxical wholeness of knowing …" by which we would "understand that genuine knowing comes out of a healthy dance between the objective and the subjective, between the analytic and the integrative …." They outline and help to teach practical ways to reform higher education to better acknowledge wholeness, and they explain the benefits to society and environment they predict would follow. Capra and Luisi (2014) in their book, *The Systems View of Life*, propose a "unifying vision" and new science concepts very similar to our own—with wholeness and systems thinking they, too, see alarming trends now as "just different facets of one single crisis," and that this is driven by "a crisis of perception" linked to "the concepts of an outdated worldview."

In addition to ideas, concepts, and ethics, holistic and synthetic science also entails and requires practical new methods of a fuller accounting. These are needed to prevent misunderstanding or illusion that proximate successes or gains are positive on net balance when all direct and indirect effects are considered (or, if not all effects, as many as can realistically be quantified and evaluated, and with special attention to impacts on Life, environment and Life-support systems). Like any new technology or tool, a true anticipatory, predictive, holistic science could be abused to manipulate outcomes for selfish or greedy ends, but that is always the case when new knowledge abounds. We reserve deeper examination of this topic for the next principle where we discuss balancing externalities with a science that inherently internalizes its own impacts and role in the world.

Principle 4. Holistic Life science equally emphasizes internalist self-referential and externalist objective perspectives.

Another quality of the science we see as needed serves to balance negative side effects of *objectivism*, which we may consider as the well-intended attempt to observe a system of study from outside it, without influencing it, so as to understand it as free from any interference on the part of the scientist. This general principle of objective science, usually taken to be fully synonymous with "good science" in the sense of rigorous, repeatable, generally applicable, and valid science, we see to be much like analysis. Objectivity is excellent when used in moderation and when counter-balanced by one or more alternative perspectives, but it becomes harmful and even pathological when used as the sole form of science or when treated as an absolute, a singular criterion, an unqualified "truth."

In many if not most mainstream science arenas, if someone's science is criticized as being "not objective," this is equated with subjectivity and bias, combining to make the motives of the scientist suspicious and results of the study invalid. However, beyond cases of actual deceit, intentional bias, and dishonesty, these seemingly solid assumptions about equating objectivity and good science bear greater scrutiny.

Like many of the sciences and concepts we examine in this book, and many others now in vogue such as principles and methods of economics, medicine and others, the formative years of the sciences and fields have occurred during an era and set of circumstances in our planetary environment that are no longer effective descriptors of current circumstances. This mismatch between the period of science or scholarly development and the conditions of the current moment also align with fields that have developed based on *fundamental working assumptions*, which are no longer true. Perhaps, the best illustrative metaphor for this is the abrupt change from real world conditions and linked assumptions of an "empty world" to those of a "full world" as described by Goodland and Daly (1996). They posed this metaphor mostly with respect to economics, but the pattern applies to other sciences as well.

During the era Goodland and Daly (1996) generalized as the "empty world," the size of human populations and the impact of human resource extraction and environmental pollution were small compared to the scale of the planet, and one could reasonably assume minimal human impact. During these many years, roughly all human history prior to the Industrial Revolution, for those developing science, economics, and other disciplines, natural resources could reasonably be treated as if infinite, and the same assumed for the waste absorbing capacity of the planet's atmosphere, land, and waters. Not only that, but it was the human resources that were scarce and the central and sole focus of conservation. Total human population, which now stands at over 7.5 billion, is estimated to have first reached 1 billion around the year 1800. However, in a relatively short period of time, these conditions, and the linked assumptions like those built into economics, changed qualitatively (and quantitatively). In the century from 1850s to 1950s (again treated approximately), the numbers of humans, the dent we collectively began to put on natural resources, and our clear footprint due to our wastes, all changed fundamentally leading to conditions better seen as a "full world." This abrupt and qualitative change in real world circumstances meant that key working assumptions about humans, resources, wastes, and the environment became invalid nearly over night, roughly about the dawn of the 20th century. This change was not simply a step function from one state to another but a fast-growing trajectory that continues to this day, thus motivating some to call this current time the age of acceleration.

Herman Daly and others used the realizations contained in the empty and full world metaphors, and much other evidence and insight, to create ecological economics—one of many new scientific fields that had to be transformed based on awareness of the profound change in the human—environment relationship and its implications for old and outdated working assumptions. One of many pathological side effects of the old form of economics (as also manifest in culture, policy, government, business, and everyday life) was the proliferation of "externalities"—impacts of business, government, and other human enterprise yielding unintended consequences to essential environmental and social capacities. These impacts are largely negative causing harm, damage, and depletion of natural resources (although positive externalities are also possible); yet, these negative impacts were ignored since they were assumed to be external to valid accounting and forecasting practices. As above, we see this as an abuse of reductionism and analysis spanning science and culture. Internalism, self-reference, and intersubjective perspectives can serve to remedy this problem and prevent others of similarly negative impact.

We describe the principle of internalism to balance objectivism from two angles. We first describe internalism in primarily scientific, conceptual, theoretical, and philosophical terms; this section is primarily information from past work. Second, we examine internalism from applied, social, ethical, and environmental perspectives; this section focuses on what we choose and intend for the future of holistic Life science. This division is not clear-cut, and the concepts and linked actions are often described at the same time.

Internalism and the relative importance of its complement, externalism (similar to objectivism), have a rich literature in the philosophy of science and related fields. Van de Vijver (1998) gave an excellent critical analysis and history of internalism that provided major concepts we need for our project in this book. She described one origin of internalism based on the work of Heinz Von Foerster in cybernetics and especially second-order cybernetics. This history and idea development was parallel to and overlapping with work being done by Van de Vijver, Salthe (2001), and others seeking new theoretical foundations for understanding living beings as self-organizing systems. Von Foerster worked during the beginning of cybernetics (first-order cybernetics) as the field "aimed at modelling purposeful behavior" in living beings and in machines (Van de Vijver, 1998). Van de Vijver wrote that modeling in cybernetics:

> ... did so in terms of control and communication, that is in terms of *external descriptions* developed on an a priori basis and implemented in one way or another in the machine. Cybernetics of the first order was a theory of the observed systems. The major dissatisfaction with this approach was the impossibility to model genuinely autonomous systems, prototypical examples of which can be found in the biological realm. The main question is indeed: how to model systems that develop their own goals themselves, that are apparently organized from within, that self-organize? (p. 297).

She goes on to recount how Von Foerster was intent to understand biological systems and how they create and choose goals for themselves. She wrote:

> In comparison to the externalist approach of first order cybernetics, one can readily call this an *internalist approach*. The attention indeed shifts from external descriptions in view of control, to questions of 'self': self-organization, self-description, internal development.

This interesting dilemma arose in and between fields seeking to understand the inner essence of living things, and to model and then mimic such internal capacities for self-knowledge and self-determination in the early computers, robots, and forms of artificial intelligence (AI) beginning to be developed. To dig deeper into these concepts requires serious commitment to complex, abstract, and challenging ideas and terminology. As we see in current debates related to AI (Bostrom and Yudkowsky, 2014), these challenges continue and are likely to be relevant and important for a long time.

Once Von Foerster and others considered the possibility of an internalist approach, more challenges arose. Van de Vijver (1998) reported some of the major questions, such as the following:

1. How do we precisely describe and define the internalist approach?
2. Does internalism imply or require "the abandonment of any form of control?" Or more generally, what happens to "control" as the relationship changes from the externalist approach of first-order cybernetics and its external model and primary role of the external engineer and programmer, to the new internalist approach seeking to work to understand "from the inside" of the system, whether living or machine?
3. Kant, Pask, and others grappled with uncertainty and the inability to know about the internal model of any system. Van de Vijver wrote: "To Kant, we will never be able to objectively know internal teleological forms; their internal circular causality will never be describable in terms of a priori principles, hence we have to add meaning in order to make sense of them."
4. What are the implications for the "relational property" unique to internalism. She wrote: "Self-organization is a relational property that attributes to the observed system the capability to observe the one who observes, to interpret the one who interprets." This reflexive or two-way observation and modeling relation was echoed by Von Foerster who said "... we have to think of a cybernetics of cybernetics ... a second order of cybernetics, a kind of self-application of the notion."

Van de Vijver (1998) and also Salthe (2001) have helped to explore and discuss these issues in the literature including studies of evolutionary systems theory. Both authors discuss important overlap with semiotics. Salthe's (2001) brief overview of internalism shares our sense that it is an essential approach we must amplify and build into the foundations of holistic Life science. He wrote of internalism:

> From the viewpoint of modern science, this is certainly among the most radical perspectives emerging at the end of the Twentieth Century. Internalism is poised over against externalism, which is just science as it has been, where the theoretician constructs a model of some part of the world as if (s)he were looking at it from the outside, therefore objectively. Such models are known to be partial... and are focused only upon aspects of the world that may be viewed for practical purposes as being mechanistic.

And later:

> There have been several attempts to move away from models of this kind in the direction of internalism as a response to the failure of externalist models when placed up against the complexity of the actual world and our own material situation in it.

Salthe (2001) related several famous examples where internalism arose largely due to necessity from the type of science being conducted. From the "Copenhagen interpretation of microphysics" emerged the "complementarity of particulate and wave/field representations of electrons" based on the realization that observer (scientist) and experimental system of study "are entangled." This entanglement thus called in to question the existence of external versus internal stances for representation, modeling, measurement, and understanding. Cosmology also led to "internalist realizations" stimulated by vast difference in scale of the observed system and the scientist-observer. As Salthe wrote: "Here we are clearly inside the system we are observing, and this has some major consequences . . . ."

These historical and conceptual topics provide a review of the ideas of internalism, how and why they have developed, and how they can be relevant and useful for a new holistic Life science with intentional founding goals to serve Life, to reverse damage to Earth's Life-support systems, and to achieve true human—environmental sustainability. Much as with cosmology, we are clearly inside the system we are observing when we study Life on Earth. Similar to the physics of Neils Bohr and others in the Copenhagen school, we can interpret our integral relationship with Life and the environment as a very real form of entanglement, in which the outcomes and implications of our experiments and science can vary profoundly based on the choices we make in formulating questions, experimental designs, analysis/synthesis, and associated activities.

One last set of quotes from Salthe, Van de Vijver, and Von Foerster on concepts help as we bridge next to the implications of internalism for responsibility, ethics, and action. Salthe (2001) wrote of similarities between internalism and dialectics:

> Dialectics also contests the objective stance of Western science, suggesting instead that investigators are (and should be) not only observers, but actors in their own interests at the same time.

Describing Von Foerster and allies, Van de Vijver (1998) wrote:

> . . . the encounter with self-organizing systems means the experience of the impossibility to describe and explain the internal dynamics of systems from an external viewpoint. This impossibility clearly implies for them an anthropomorphic as well as an ethical move: (i) we have to assume that those systems are able to interpret our interpretation, and hence (ii) our own behavior, our own choices, values and decisions have an essential place in the theory of self-organizing systems. No objective knowledge is at stake: it is the interaction in view of certain goals, like consistency, like survival. It is not appropriate to call this an externalist position, as the objectives of control, and the purported adequacy between internal and external are abandoned (p. 299)

Note her mention of "interaction in view of certain goals . . . like survival." This insight gets right to the root of our present need—a science of the human interaction with all Life in view of the shared goal of survival. She also quoted Von Foerster on why internalism is radically different, first speaking about the classical or externalist approach:

> You speak about something else . . . And at the moment the separation between you and what you are saying is made, my feeling is that any notion about

ethics and responsibility is already subdued, suppressed … You don't need to be responsible if you are only speaking *about* something.

And then describing the internalist alternative:

It is not going that nice classical way any longer … Whatever you say, it is *you* who is saying it … at the moment you speak about you then it is *you* who is speaking and therefore you are responsible. (p. 296)

Van de Vijver (1998) notes the implications:

What a bold statement this is! Can scientists ever get seriously involved with such an idea? … it implies that there is no room for hiding, no escape of the speaker is possible: everything you say, you said it, and it has to be taken into account as such. (p. 296)

She goes on to describe how this is "revolutionary" in relation to most of philosophy, including Kant, who assumed the "existence of objective knowledge" and worked from there, but "Kant is speaking about the possibility of objective knowledge; he is not speaking about himself."

We seek to follow this power of internalism to provide a self-referentially consistent science. We seek concepts and modeling that make sense and provide actionable intelligence both when applied in a first-order sense (an ecology of Nature, or an environmental science of the world) *and* in a second-order sense (an ecology of ecological science, an environmental science of environmental science). We embrace the imperative to consciously accept responsibility for our science, words, and actions—to speak about ourselves and the world at the same time—and see internalism to help make this responsibility *explicit and transparent*. Operating under mainstream externalism and objectivity, in reality, we still are responsible for the effects of what our science does to the world and to ourselves (for the ultimate outcomes on Life-support systems, other species, and people of the future) of both daily science operations and the manifold downstream repercussions of the data, results, methods, models, and conclusions we publish. But, perhaps constrained by an artificial and overly narrow view afforded by extreme objectivity, we may ignore this responsibility, treat it as "external" to our science, find excuses for leaving it outside the system of study, and fail to incorporate such ethics into our founding ideas and institutional transmission of our fields and disciplines.

This kind of fragmentation between internal and external theories can lead to academic institutions receiving millions of dollars in government and other grants with stated goals to help understand and solve the sustainability crisis, while operating within buildings, and with technology and habits, that have nearly the same negative impact on atmosphere, water, species, and energy (thus, making worse key global ecological and social symptoms) as the industries and practices usually treated as causes of the problems—energy, transportation, agriculture, mining, entertainment, manufacturing, etc.

As we develop a "second-order ecology," an ecology of ecology like von Foerster's cybernetics of cybernetics, we turn the "ecoscope" inward in addition to outward. We seek modeling and science general and robust enough for us to apply to ourselves at same time as we apply this holistic Life science to the world. We are also able to apply this science to the relationship between our science and the planetary system in which we are intimately embedded. A large proportion of the dynamic systems we study and for which we seek to build a science and theory are self-organizing, alive, and thus also observing, interpreting, and responding to us as we observe them. If we adopt the humility and responsibility inherent with internalism—even if we use it only half the time as an equal complement to mainstream externalism and objectivity—then this fundamental stance leads to new imperatives and promises a new quality of results, impacts, and outcomes.

We see a central aspect of holistic Life science like Van de Vijver described internalism—"attention shifts from external descriptions in view of control, to questions of 'self': self-organization, self-description, internal development." In harmony with our Chapter 2 and Principle 1 in this chapter, this attention shift leads to a shift in values. As we give up on control of other living systems, as we acknowledge and respect the unknowable internal self, life, and right to exist in other Life, we in essence elevate the value of those living and environmental systems to equal status with ourselves. Related expressions of this perspective are the following: (1) the move to make Life the highest value has the effect to put all Life, including humans and the environment, on the same value level; (2) we have a better chance of survival if we identify, cooperate, and ally with all the other Life systems who share this survival goal; and (3) if we accept the concept of sustained Life as integrated with ecosystem and biosphere scales of Life organization, then we humans are always "inside the system of study" and participating parts in a larger living whole that is also a larger living self with its own internal models and unique interior realm.

As we look closely at the world now, we become aware that we are seeing our own reflection. The symptoms of crisis we see are our own doing, not only the results of our numbers, actions, and technology but also the direct results of our science, values, systems of ideas, and ways of thinking. This environmental awareness and attendant self-awareness can be a stimulus to develop a more self-reflective science. The next step is to realize that we need ways to bring humans, scientists, and the science process *inside* the science paradigm itself, such that there is a consistency and closure of doing, practicing, and objectives. And, much as our grand goal is science to help sustain Life, we also seek a science paradigm able to sustain science by serving to value and maintain the essential context and Life-support system (science support system) with which science, like any living system, must coexist and coevolve in win—win mutual relation.

Taking as given that we love and value science greatly, we must help science care for its own legacy and its own health. Moreover, we must be holistic and comprehensive to understand and care for the legacy and health of science in all forms—not just narrow subsets of the science enterprise and output, like students taught and advised, publications, grants, etc. We also need to internalize care for the facilities, operations, ability to continue and sustain, and our impacts on the planetary environment.

One way to achieve this is to develop and set goals, or adopt existing goals, for transforming our science facilities, schools, labs, and field sites to be environmentally sustainable. Many universities have done this via the American College and University Presidents' Climate Commitment (ACUPCC, 2018), Sustainability Tracking, Assessment, and Rating System (STARS) of the Association for the Advancement of Sustainability in Higher Education (AASHE, AASHE STARS, 2018) metrics, and other targets and transition plans. These commitments and cooperative monitoring and change programs can help to identify targets and trends for the carbon, energy, water, nitrogen and other ecosystem fluxes as we move toward sustainable practices. Some organizations and even governments set target dates, such as to achieve climate neutrality by 2020, 2030, or 2050. Great works are readily available providing still more key indicators that can guide plans and change—see for example the planetary boundaries (Rockström et al., 2009), ecological footprint (Wackernagel et al., 2002), Genuine Progress Indicator (GPI, 2018), and Millennium Ecosystem Assessment (MEA, 2005). To incorporate internalism fully, to internalize all those environmental impacts previously ignored and externalized, these kinds of systemic goals must be met; and, we must begin immediately and keep track continually of the trajectory to success. Given what we know now, we have no excuses, no exits—as internalism requires, both our words and our actions speak about ourselves as well as our world, and there is nowhere to hide.

This approach may seem to cross the line into bias, a bad form of subjectivity, or activism that threatens to tarnish the public trust in science products and process. We assert strongly that this need not be true, and that the evidence, rational case, and ethical imperative of the full necessity of valuing, understanding, and sustaining Life eclipses any such doubts. Furthermore, if we seek trust from the public so as to be able to teach, inform, and help guide, then we would benefit from holistic ethics, from unifying thought, words, and actions, as in the words of Albert Schweitzer (Byers, 1996) who said, paraphrased:

> For influencing others, example is not the main thing. It is the *only* thing.

Thus, as we convert ourselves and our science operations to sustainable practices first, we *increase trust* as we remove hypocrisy and double standards. Students and the public can read, hear, and absorb what our buildings, labs, technology, habits, and culture teach about our *true understanding and respect* (or lack thereof) for Life and environment. Moreover, these actions can "speak louder" than the words in our lectures, textbooks, academic exercises, and exams. We can practice what we preach and teach; we can use our R&D capacity and expertise to develop the science and applications of sustainability to be sustainable themselves, at the same time and in mutually beneficial synergy.

As a segue to the next section on complexity and the need for multi-scale methods, we outline briefly a hypothetical process to help scale up, scan outward in space and forward in time, to gage the impacts of any science theory or action. This table starts at a local, small, personal scale of the scientist and provides three stops on the way to a wider assessment of impacts.

Entity or Holon	Spatial Scale	Time Scale	Social Units
Organism	0.1–1 m	1 s to 1 day	Individuals, team
Ecosystem	100–1000 m	1 week to 1 month	Community, region
Biosphere	1 –10 km	1 year to 1 century	Nation, humanity

This kind of bridgework or scaffolding, even if done quickly and qualitatively, or via thought experiment, can help to span the boundaries between narrow and artificially isolated subsets of the world that we simplify in order to study, to integrate to more whole and complex systems in their authentic interdependency. In a general way, to evaluate the extended impacts of any science idea, theory, study, or activity would be like the tradition in some Native American cultures of considering the impact of decisions on the seventh generation in the future (Kirmayer et al., 2011). We see that the concept and necessity of a long-term and intergenerational perspective and equity is not new. Then why is it so hard to implement? Specifically, regarding the role that we control and modify the environment, especially without an overriding respect for the value of Life, we must be cognizant of the lasting extension of our impacts, which create path-dependent futures, thus dictating opportunities for those yet to follow. Author C.S. Lewis expressed a similar concern in his work, *The Abolition of Man*, in 1959:

> In order to understand fully what Man's power over Nature, and therefore the power of some men over other men, really means, we must picture the race extended in time from the date of its emergence to that of its extinction. Each generation exercises power over its successors: and each, insofar as it modifies the environment bequeathed to it and rebels against tradition, resists and limits the power of its predecessors. This modifies the picture which is sometimes painted of a progressive emancipation from tradition and a progressive control of natural processes resulting in a continual increase of human power . . .

This process of long-term thinking also aligns with more holistic and comprehensive accounting processes, such as life-cycle assessment (e.g., Guinée, 2006) and triple bottom line methods (e.g., Slaper and Hall, 2011), but they are not employed in standard and ubiquitous fashion.

The next section on complexity also deals with a science able to reconcile seeming opposites and accept apparent contradictions, and so we can admit here that the plea just above for a sustainable science of sustainability is rooted in the Sustainer worldview. As such, we also realize other scientists, perhaps operating from and espousing a Transcender worldview (see Chapter 1), may disagree totally with our assessment and recommendation. This is to be expected, and rather than seek to defeat or refute the differences in views with Transcenders, we seek to unite and cooperate.

> **Principle 5.** Holistic Life science is complex itself and is able to model, understand, and recommend wise actions for interacting with and sustaining complex systems in Nature.

We use the term complex here primarily in the sense of Robert Rosen (1977, 1991, 2000). As for the other five principles in this chapter, we see complexity as an essential founding principle that is not currently understood, appreciated, or employed fully. Rosen's first work on complexity focused on the relational aspect—how complexity arises in one's interaction with, or study of, any system. He wrote (Rosen, 1977):

> Complexity is generally viewed as an intrinsic property of certain kinds of systems, or at least, as a property of a specific description of such systems. The view towards complexity taken in the present note is different; namely, that complexity reflects the necessity for many distinct modes of description of a system. This in turn depends upon the number of ways we can effectively interact with a system, and ultimately on the number of distinct subsystems which available observational techniques make accessible to us.

In other words, using this approach, complexity cannot be measured as an independent state variable, but rather in context with its environment. Even some of the approaches that use thermodynamic principles (energy storage, exergy storage, energy throughflow, retention time, cycling, etc.—see Fath et al., 2001) are measured relative to some reference state. A far from equilibrium system has an absolute equilibrium, in terms of universal background temperature, but more relevantly a local, contextual, and transitory environmental reference. How much work can be extracted from a system (one measure of complexity) depends on this local situation. If it were simply an intrinsic property, then that system could be isolated from its environment and reduced to its constituent parts. The number and type of parts and subsystems would be an indication of the complexity.

In later works, Rosen's definitions and uses of the terms "complex" and "complexity" both seemed to increase in complexity! Following his own definition above, at least in the sense of using language as a means of modeling, he used "many distinct modes of description" of complexity, although all are interrelated. At various times, he referred to complexity (Rosen, 1991, 2000) with respect to these main ideas, which we have classified into groups:

1. Necessity of multiple distinct modes of description and interaction (as above). This relates to his idea that a complex system has no single model capable of representing the system.
2. Complex systems are beyond formalization, simulability, and computability. This is similar to his syntactic versus semantic distinction. In addition, he

points to Gödel's Incompleteness Theorems as proof of this issue using the example of inherent incompleteness of formalizations in mathematics.

3. Generic (complex systems) versus limited or special cases (simple systems). He asserts that complexity is the norm and generic case whereas simple systems are rare and artificially constrained cases.

4. Simplicity and simple systems—to describe the opposite of complex and complex systems. This is similar to his references to mechanisms, machines, and mechanistic approaches as only applicable to simple systems and not complex systems. We see this as also overlapping with his statements on fractionability—complex systems are unfractionable.

5. Entailment as a more general concept that has two major types—causal entailment in natural systems in the real world and inferential entailment in models and formal systems. We take Rosen's use of entail to mean the same as the dictionary definition: "to impose, involve, or imply as a necessary accompaniment or result" (merriam-webster.com). Entailment is needed in Rosen's "modeling relation," which he saw as the essence of the scientific process. We align the main points of this book with need for a revised modeling relation much the same as we have proposed the need for a new paradigm in science and new system of ideas in culture. Rosen also employed entailment differently for complex systems (and life) compared to simple systems (and machines).

6. Impredicativity—he wrote of the need for impredicative logic to understand and model complex systems such as living organisms.

We will address most of these, as the set of interrelated facets of complexity has value for our project to better under Life and living systems, for interactions with them, and for interventions seeking to steer complex human–environment systems toward sustainability, such as via science and technology. Our ultimate claim in this section, which we work to develop and support, is that Rosen's closely integrated distinctions between (1) complex and simple systems and (2) living systems and machines are crucial for both understanding the causes of our current human–environment crisis and for developing a new holistic Life science, and technologies based on it, to solve the crisis.

A series of diagrams may help to clarify these many interwoven, and yes—complex—ideas about complexity. The first two diagrams (Fig. 5.3) are reproduced as slightly modified from Rosen (2000). The third one he described in his text but did not actually present that graphic (Fig. 5.4).

Rosen used these two figures to differentiate the mainstream and prior science views on the left (his Figure 19.1) from something closer to his own theoretical perspective on the right (his Figure 19.2). He spoke of classical science dominated by physics that assumed that organisms were nothing more than special case subsets of machines or mechanisms. He also wrote of von Neumann and others who believed that machines could achieve life just as organisms do, while Rosen himself treated organisms as categorically different. Complexity

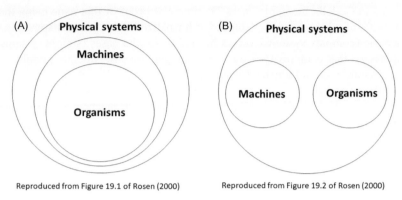

Reproduced from Figure 19.1 of Rosen (2000)   Reproduced from Figure 19.2 of Rosen (2000)

FIGURE 5.3

(A and B) Rosen's conceptual diagrams for machines, organisms, and how complexity is essential for understanding both.

determines the boundary and threshold between organisms and machines in his Figure 19.2. He did not publish Fig. 5.4, but he did describe it (Rosen, 2000):

> ... I suggest a taxonomy for natural systems that is profoundly different from that of Figure 19.1 or Figure 19.2. The nature of science itself (and the character of technologies based on sciences) depends heavily on whether the world is like Figure 19.2 or like this new taxonomy.

> In this new taxonomy there is a partition between mechanisms and nonmechanisms. Let us compare its complexity threshold with that of Figure 19.2. In Figure 19.2, the threshold is porous; it can be crossed from *either direction*, by simply repeating a single rote (syntactic) operation sufficiently often ...

> In the new taxonomy, on the other hand, the barrier between simple and complex is not porous; it cannot be crossed at all in the direction from simple to complex; even the opposite direction is difficult. (p. 293)

Here, we note that Rosen is equating or associating mechanism as simple and organism as complex.

Impredicativity is another facet of complexity that Rosen studied and showed to be fundamental to understanding Life. It relates closely to our founding principle of holism above. Rosen (2000) wrote that something is impredicative if

> ... it could be defined only in terms of a totality to which it itself had to belong. This ... creates a *circularity*: what is to be defined could be defined only in terms of a totality, which itself could not be defined until that element was specified. (p. 294)

He goes on to say that formalizations and simple systems cannot include any impredicativities or forms of self-reference, while complex systems and models

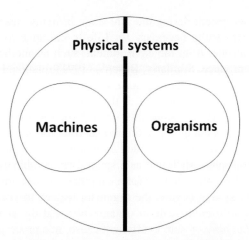

Modified from Figure 19.2 of Rosen (2000) and as
suggested by his mention of a "nonporous partition"
between mechanisms and nonmechanisms

FIGURE 5.4

A depiction of the "new taxonomy" for natural systems Rosen described in which simple machines and complex organisms are categorically different types of systems requiring fundamentally different science and technology.

of them must include impredicative aspects. This connection to self-reference corroborates the need for our principle of internalism above, which is also aligned with self-reference.

Rosen employed these multiple conceptual facets of complexity, along with category theory, system modeling, and related mathematics, to develop his unique strategy and answer to the question, What is Life? He wrote (Rosen, 1991):

> ... *a material system is an organism if and only if it is closed to efficient causation*. That is, if *f* is any component of such a system, the question "why *f*?" has an answer within the system, which corresponds to the category of efficient cause of *f*. (p. 244)

Rosen further explained this provocative result and its implications and wrote that machines and mechanisms, and their associated models, suffer from an "impoverishment of entailment" as compared with living systems such as organisms.

Despite the elegance, potentially generic validity, and seemingly revolutionary impact of Rosen's ideas, as with any great work, he also serves to inspire many new questions. While Rosen focused his study and modeling of life on the organism—which we treat as just one of three holistic Life unit-models, and which must be integrated with the ecosystem and biosphere—his closure to efficient cause faces another opening when we consider the making of an organism beyond any individual life span. That is, while the organism-maker function may well be inside the organism-as-system during one life span, in many cases to make a next generation of organisms (fully necessary for Life to continue to exist long term) requires something outside, from the environment, such as another organism with which to mate and reproduce. Similarly, Rosen's (1991) relational model of organism includes a representation of inputs from the environment that are transformed during metabolism in his focal function f. If these inputs are available at a rate that ever drops below the necessary input rate needed for metabolism, then Rosen's organism faces another opening for a functional role for something outside the organism. We see the need for two coupled complementary types of organisms—autotrophs and heterotrophs—such that as a synergistic combination they are able to recycle the essential resources inputs for each other and thus embody "sustained life." This larger Life system beyond the organism we see to also reflect Rosen's key insight, albeit at a different scale of organization—at the ecosystem and biosphere scales, we again have closure to efficient cause, in a process of "self-making of the self-makers" which also helps to explain and enable sustained Life.

Much like the impredicative, self-referential loops Rosen invokes—like holding two mirrors to face each other and looking down into the indefinitely repeating reflecting images—we can and will dig even deeper into Rosen's contributed complex wellspring of insights. However, we save further discussions of his ideas for Chapter 7, where we examine the bridge between science and real-world applications and Chapter 8, where we look at examples of holistic technology and applications. There we utilize his modeling relation, in which the key achievement is a commutative property such that inferential entailment in a model faithfully represents the causal entailment in some natural system. We also use ideas of Rosen's warning of infinite regress when adding control loops to simple models and mechanistic systems. Also, later in the book, we examine why it matters to keep machines and Life categorically separate.

Principle 6. Holistic Life science is radically empirical with constant capacity for questioning, challenging, and transforming ingrained assumptions and structures quickly and efficiently.

This principle is necessary to recognize formal and founding value for a culture and practice of science that has a robust means of self-critique to keep itself honest and focused on Life value. When holistic Life science, and the scientists practicing it, are unafraid, encouraged, and rewarded for saying what is actually happening, that culture of science will have the capacities to (1) resist peer pressure to conform or to avoid certain types of questions, such as challenging questions related to sustainability; (2) resist pressure to remain within paradigm and continue the normal science program of "puzzle solving" (Kuhn, 1962); and (3) make questions and forays into "post–normal science" (Funtowicz and Ravetz, 1993) always available and valid. These capacities for staying focused on what is really happening are necessary to resist any social, economic, or other pressures that could distract scientific attention and resources from those topics of greatest benefit to humanity and Life.

The allegory or metaphor for this principle is the story of The Emperor's New Clothes. In this fable, while all the adults play along with the delusion and charade of the King and the salesman, it is left to a child to be the courageous and radically empirical one to speak the truth. In a similar way, we propose this principle (which we hope can be developed, refined, and strengthened over time) to help holistic Life science retain a similar value system in which truth—represented by those truths anchored in Life value—has greater power and authority than social or economic forms of peer pressure.

This radical empiricism is made more possible since we have anchored and grounded holistic Life science on the value of Life. The courage to speak truth, Life value truth, can be bolstered by solid foundations when tests and critiques are made relative to radical respect for Life itself. Here, we employ one meaning of the term "radical," which relates to getting to the root of a matter. When the values of science are anchored to a deep basis that is perennial, timeless, and unchanging, it can help individuals and institutions steer a steadier course and not be blown off track by changing winds of political, social, or economic "hot topics," norms, or fads.

We will describe just two examples of what radical empiricism would look like in practice in this chapter and then propose possible future efforts in this vein. One necessary capacity of radical empiricism is the ability to test our deepest assumptions, and we see the need to do so regularly. These tests would best be done by actual physical experiments; however, thought experiments can serve this capacity at least for an initial form of testing assumptions. For example, we see one such deep assumption in the current mainstream science model of "life as organism" which is linked to the assumption that "life is separate from environment" (or life is other than environment, and thus can be valued less than life). We can test these assumptions, and our having done so has informed the work in this book. What we see as the result of a test of these assumptions, is that when we *actually* separate life from environment, life is destroyed (using life here instead of Life to fit with the existing mainstream paradigm).

A physical implementation of this experimental test would be to isolate any individual life form (again, a discrete life form, or organismal life form, as in the current dictionary definition of life) in a closed container, like a glass jar or chamber. Whether this individual life form is an autotrophic plant, a heterotrophic animal, a microbe, a human, or any other type, it would not be able to live for long. The only intermediate result would be a delaying tactic of some life forms which might be able to go dormant and wait for the environment to change. But, in general, by actualizing a real system (by implementing such a system in experimental form) in which "life is separated from environment," we must confront the radical truths that (1) an organism or individual is an incomplete instance or unit-model of life and (2) life and environment are inherently inseparable and unfractionable. This leads to the further realization that it is not enough to have a single model of Life (like "life = organism"). If we could use a single model, then it would mean that Life is simple and has a single reducible model. As we saw in Rosen's work on complexity above, this assumption does not fit well with either evidence or logic.

Another example comes from an educational experience. In a seminar course in graduate school, we read the famous Likens et al. (1970) paper on their experiments in Hubbard Brook forested watershed ecosystems. In this enormous experiment, they clear-cut a forested watershed and then used herbicide to prevent vegetation regrowth. By studying the minerals and essential Life system nutrients that were rapidly exported via the stream (e.g., nitrogen, calcium, magnesium, and potassium), they learned about the natural homeostatic properties of forested ecosystems and what happens when they are disturbed. In the conclusion of this seminal paper, they wrote that homeostasis and a continued healthy ecosystem depend on "a functional balance within the intrasystem cycle of the ecosystem," and they referenced balance between production and consumption via decomposers and nutrient cycling. This experiment was of profound historical, conceptual and scientific importance and had been read and studied by all the leaders in the university environmental science institution in which we were enrolled in a PhD program. However, looking around the room where the graduate seminar was meeting, it was immediately clear these core lessons from Likens et al. (1970) had not been learned in any deep way or in a radically empirical way. The lessons of the fundamental importance of functional balance and intrasystem cycling were not employed in the design of the academic science building, and they were not embodied or operational in the science or educational enterprise and its associated apparatus. It was as if those hard-won lessons of that huge experiment were treated as for "academic purposes" only—interesting topics to read, and discuss, and perhaps even employ in other scientific studies, but not grasped as fully relevant to the Life, culture, practices and operations of ecological and environmental science itself.

This seeming disconnect between lessons learned from Nature and how we live and act in academia has changed greatly since this graduate seminar. Many now follow the early work by David Orr at Oberlin College (Orr, 1991) and his insistence that we use university campuses as a laboratory and for experiments to transform our universities to sustainable operations. This leadership was followed by the American College and University Presidents' Climate Commitment (ACUPCC) and related efforts for radical change for sustainability in colleges and universities. Many other self-change leaders have added works and voices to this chorus, including Chris Uhl at Penn State and John Aber at New Hampshire. Here, we see another tie back to our principle of internalism, another example of the consistency of the six principles.

These academic leaders were not usually rewarded within the existing system of academic advancement and recognition. Many times, they had to buck the system and go against the grain, and most often did not risk such radical work before achieving tenure due to the professional risk. Empiricism and the principles, ethics, habits, culture, and practices of Life science should not be biased, bound, or constrained by money, pay, rewards, peer pressure, ego, tenure, or stature.

Socioeconomic pressures and careerism should take a backseat to a science enterprise oriented and pragmatically able to deliver the best for humanity, including elevation of the value of Life and Life support to its rightful place as #1 priority that must be ensured for other human endeavors to have meaning or even be doable into the future.

If Life value could be institutionalized into the rewards and value structures of academic, government, and private practices of Life science, then this principle would enable accelerated learning, change, and paradigm shifts when the criteria for success and value are grounded in what is needed to support Life. Sidetracks, distractions, and coopting of the science enterprise for work that harms Life would be minimized or eliminated. Thus, another benefit would be more efficient use of limited science resources, such as funding and people power.

SUMMARY OF THE SIX PRINCIPLES

This set of science principles embodies an ethical sense that emerges from the value orientation to serve Life, from the anticipatory program to succeed in achieving sustainability, and from the other principles that promote leading by example and applying the holistic Life science to the scientific enterprise itself. This ethical imperative is similar to the Hippocratic Oath taken by medical doctors—*Do No Harm*. These principles also hold capacity to create a science that will be seen by future generations in a positive light—we can be more confident that future scientists and citizens will look back on our generation and see us as good ancestors, elders, forerunners, or grandparents. Much like the Native traditions we referred to above, we see holistic, anticipatory, self-referential Life science to provide a solid basis for such generosity of forethought and service to seven or more generations in the future. If we cannot ascertain or be confident that a scientific or technological program or project will be net positive for Life and the future, then we should not do it. Just as with conflict of interest, if the motives and rewards are not clear and well founded, then the initiative should not be trusted and should not be conducted. With such high standards holistic Life science can be a platform for us as servant leaders.

These principles also hold potential to enable greater scientific insight and understanding and to open new avenues for holistic action for sustainability. One example is application to modern, industrial human systems in the case of the US food system. Multiple studies have shown that our current food system uses approximately 10 units of fossil fuel energy for every 1 unit of food calorie energy delivered to people via their diet. This network-level synthesis view is only visible via whole system analysis, whole network *analysis plus synthesis*. This analysis + synthesis, when combined with radical empiricism and the other principles, indicates clearly that the US food system is not actually effective, functional, or efficient for delivering food (as measured in energy for human metabolic needs): clearly, this approach to agriculture is not sustainable when those inputs are nonrenewable fossil energies and the outputs burden the atmospheric balance. By understanding Life, and the necessity of its relation to environment and energy, we know that no rational food energy system would expend 10 times the energy it supplies. In search for explanation, it becomes clear that the food system is designed for some other primary purpose and that the grounding to Life value has been lost due to distraction, confusion, or other negative influences. As we examine further, and as we observe that the food system does function for generating financial profit for some individuals and corporations involved, we are confronted with the need for fuller accounting to internalize the many social and environmental costs now being externalized. Employing radical empiricism and the other proposed principles of Life science, we must address such hard reality and develop solutions for systemic change in light of ultimate values grounded in Life.

Ecological network analysis and systems ecology provide many holistic and systemic ideas, methods, case studies, and results that serve as an excellent foundation tool kit for the Life science we propose. We next discuss seven major lessons from these holistic fields and how they can be employed.

Life science lessons from ecological networks and systems ecology

6

In this chapter, we present transformative lessons from ecological network analysis (ENA), systems ecology, and related fields, including analytical tools, science results, case studies, models and their validation, holistic hypotheses and their tests, and an equation for Life—environment using hypersets. Our central goal is to identify tools able to translate from the values and principles of holistic Life science (Chapters 2–5) through to the proximate goals of applications, technologies, and solutions (Chapter 7 and Chapter 8) and on to successful outcomes and the ultimate goal of a win—win human—environment relation by which the environment improves over time as a result of human influences (as framed in Chapter 1 and the focus of this book).

We outline seven key lessons from systems and network ecology that are starting points for essential concepts and tools for the new holistic Life science. These Life Lessons are as follows:

1. Discrete versus sustained life. We summarize and present in concise form all the subtopics presented in Chapters 1–3 on this topic. This includes recognition of "coupled complementary processes" (Fiscus, 2001–2002) as a bridge relationship between discrete and sustained life.

2. Ecological goal functions—developmental tendencies and the orientation of change of whole ecosystems—are complementary and can be reconciled (Fath et al., 2001). For example, maximum network dissipation and maximum network storage though opposite tendencies can both occur as long as network retention time is also increasing.

3. Indirect relations and impacts are often greater in magnitude and can be qualitatively different (positive versus negative) than direct impacts between any two entities, for example, two species in a community or food web or two firms in an economic web. This means detailed, fully articulated whole networks or systems (or as close to whole as is reasonable) must be studied. Indirect relations and impacts integrate over all the connected relations in a network (or community, ecosystem), and we cannot see this predominance of indirect relations via narrow studies, on a few isolated entities, when many interactions are truncated and ignored.

Foundations for Sustainability. DOI: https://doi.org/10.1016/B978-0-12-811460-5.00006-6

129

4. All Life is connected—via ENA, we can quantify how the connections, relations, flows of energy and material between Life forms are crucial to understanding Life. These network relations between all entities are equally as important as more traditional approaches focusing on dynamics (as in dynamic process models) and measures of materials such as the particles, objects, or masses of individual entities. This lesson adds quantitative evidence to the social and moral sense of the need for a new "declaration of interdependence" to formally recognize the organic unity of all Life.

5. Ecosystems and natural networks show mutualism and synergism between species, actors, and participants. As a general pattern, "life is beneficent under network rule" (Patten, 1991; Fath and Patten, 1998), but the details and specific methods to see this systemic quality are important.

6. Ecosystems and networks naturally balance order and flexibility. The robustness index of Ulanowicz (2009a) uses information theory to quantify the balance of efficiency versus adaptability (linked to redundancy) in ecosystems and flow networks in general, including economies.

7. A hypothetical new formalism is presented which can serve to prohibit fragmentation of life from environment and of life from life. This formalism employs hypersets—impredicative, self-referential, meaningfully ambiguous mathematical tools as suggested by Rosen.

INTRODUCTION AND BACKGROUND FOR THE SEVEN LIFE LESSONS

ENA and systems ecology provide many holistic and systemic methods and results that serve as an excellent foundational tool kit for the holistic Life science we propose and the systemic solution project we promote. These are the fields we have worked in for decades, the disciplinary tools with which we are most familiar, and those methods and results we see to be most closely integrated with the holistic Life science we have begun to develop. However, we are also aware of allied workers in related fields who employ kindred concepts and methods, who have obtained corroborating results, and who have blazed new interesting lines of investigation using network science.

We seek to remain in cooperation with these folks whether as actual collaborators or mutually informing references. We have already mentioned three people who we see as leaders in this work, scientists from whom we have borrowed and learned much. Robert Ulanowicz has developed an ecological metaphysic, most recently laid out in his book, *A Third Window* (Ulanowicz, 2009b), and he elegantly describes his perspective on a paradigm shift and new foundations for science compatible with our views. Several of the methods in ENA he invented and developed figure into our list of critical tools below. Patten (2016a) has similarly presented a comprehensive scientific framework he calls *Holoecology*. This masterful work details his 20 cardinal hypotheses for understanding causality in complex adaptive hierarchical systems. We borrow several of his cardinal hypotheses for our list of seven key lessons of Life learned from systems ecology and ENA. Sally Goerner developed a holistic approach to sustainability science (Goerner, 1999) and most recently has led efforts to bridge between ENA methods developed by Ulanowicz and others in order to apply them to economic networks. Her inspiring interdisciplinary and systems science work spans the whole of human history. She is expert at framing complex topics in readily accessible language, metaphors, and imagery, and developing pragmatic tools and metrics to aid sustainable development of vibrant human economies.

Beyond these three closest allies, we know of a large community of practice in areas akin to holistic Life science. Louie and Poli (2011) have followed up on Rosen's work helping to apply his category theory, modeling relation and impredicative systems to living systems, human psychology, and social realms. Capra and Luisi (2014) present compatible work employing systems and systems thinking as an alternative to the mainstream mechanistic paradigm. Stephan Harding and his colleagues at Schumacher College teach unique graduate programs in Holistic Science and have continued Lovelock's work after Gaia (Harding, 2006). Alan Savory, Lester Brown, Joel Salatin, the Rodale Institute, and others in regenerative agriculture and permaculture are leading the way to truly sustainable and regenerative food production methods successfully improving soils and other key elements of environmental quality as they operate.

Having gotten this far in this book, you might be aware of these milestones or branch points along the journey, or some similar series of experiences and thoughts.

You are willing to entertain our conceptual framing of the current human—environment crisis and one avenue for systemic solution—reform of science foundations to better enable understanding and actualization of the already demonstrated win—win Life—environment relation we see in most natural systems (Chapter 1).

You are open to the idea that there may be two distinct camps with respect to this crisis, and two categorically different assessments of the meaning, and entailed action plan, related to our current set of world circumstances. These two lightly held (by us) models of types of people and their often strongly held worldviews are the Sustainers and the Transcenders (Chapter 1). You have accepted our working assumptions and hypotheses for the purpose of this book that these two groups interpret and act in distinct ways when confronted with real evidence of environmental and resource limitations and negative harmful impacts of human activities. Transcenders, when confronted with limitations, interpret them as limits to be transcended, and they thus focus on ways to increase the pace and efficiency of resource extraction and utilization to fuel innovation, invention, technology, growth, and development in order to smash through any physical limitations with sheer human ingenuity, determination, and hard work. Sustainers by contrast see the current real evidence of environmental and resource limitations and negative harmful impacts of human activities as essentially insurmountable at the scale of planet Earth. Thus, when confronted with this limitation, they interpret it as a hard constraint—a system itself, environmental, ecological, human, on the other side of the barrier that deserves respect and understanding, that must be accepted, and they thus focus on ways to reorient and reinvent human activities in order to abide by the limited real capacity of planetary Life support systems.

If you identify more as a Transcender, then we have noted that this book may not provide foundational science, values, principles, methods, and applications for the "Transcender program." But, you hopefully are aware that other enterprises can be compatible with the Transcender worldview and action plan while also respecting the validity of the Sustainer approach. Efforts related to colonizing Life beyond Earth, for one example, are not only compatible with allowing Sustainers to lead the work on Earth but also reveal the essential complementarity of Sustainer and Transcender modes (how else will you sustain Life during space travel and on any space station, colony, or new world?).

Whether identifying as Sustainer or Transcender, or perhaps agnostic and still learning with respect to these hypothetical worldviews, you are open to considering Life as the primary basis of value (Chapter 2).

You are aware of the options and implications for choices of system boundaries when developing a study of real-world systems. This includes the important realization of the impacts of what one considers inside or outside of a system of study, and what aspects of the real world one chooses to separate and split apart versus those seen as unified and whole (Chapter 3).

You have considered holistic approaches to addressing those deep and wide fundamental questions of the original and fundamental nature of Life (Chapter 4).

You have read and considered six founding and interwoven principles of a potential reformed paradigm of holistic Life science, namely, (1) anchoring science on Life value, (2) anticipatory capacity to enable a successful solution to the human—environment crisis, (3) holistic emphasis to balance predominantly reductionistic current and past practice, (4) internalism emphasis to balance mainstream objectivism, (5) inherent capacity to handle complexity, and (6) radically empirical strategy that gives greater authority to Life value than to social and economic norms and forms of peer pressure (Chapter 5).

Thus, at this point, you are ready for specific, quantitative, scientific tools and techniques, pragmatic methods that are not only analytical but also synthetic. You are open to learn about innovations in science able to inform actions and enable successful outcomes where prior methods have gotten stuck, delayed, or sidetracked. You seek new ideas and strategies to aid existing science culture that still needs assistance if we are to avoid grave damage to people and planetary Life support and achieve a positive future in mutually beneficial relation to our environment and home.

Are there fundamentally different ways to formulate the scientific questions we ask that could lead to better outcomes and more rapid progress? Do we have access to different starting assumptions about how best to do science? Are the current norms and values about "good science" acceptable? Should we expect more from our research projects than new information leading to new publications? Do we trust that the stepwise and incremental accrual of new knowledge can scale up to address the systemic human—environmental crisis we are confronting? What science methods and techniques are available to help balance holism with reductionism and synthesis with analysis? These are the types of fundamental and existential questions we have been exploring for many years and for which we seek to contribute answers and insights.

The seven lessons below are the most powerful ones we have gleaned from our studies of Life in the natural world and the human relationship to it. Based on many experiences using these methods, we have developed the holistic diagnosis of a scientific and cultural "system of ideas" as the root cause of our crisis, as depicted in Fig. 1.2 (and as causing the top 11 symptoms of global environmental crisis in Table 1.1).

Given that we identify science as part of the problem, and see science as part of the solution, and are employing science to develop the solution, this is a complex, multi-perspective, self-referential process in the sense of Rosen—we attempt to pull the scientific process inside our project system as an internal function of self-transformation and to help develop a science of science. This is similar to the essential "closure to efficient cause" quality of life-as-organism that Rosen discovered. By analogy, we hope that this effort can "take on a life of its own" and mimic the self-making aspect of Life systems. By way of a more humorous if dangerous analogy, it at times feels a lot like "building the plane while flying it."

Before describing the set of seven Life Lessons and science methods, we want to make conscious starting steps and be aware of preparations largely ignored and not done now in the current dominant mainstream, reductionistic, mechanistic science that we hold partly responsible for our current crisis. These are necessary and major considerations that guide the strategy of science given our rare and specific context at this point in history. We use Rosen's modeling relation as central conceptual framework. We also consciously employ internalism and internalization, an anticipatory approach, holism, complexity, radical empiricism, and the Life value principles of Chapter 5.

First, for the context, we propose a necessary awareness that we are in a transition period. This highly dynamic era, perhaps like a transient period or even a phase change, phase transition, or bifurcation moment, allows us to recognize, accept, and even participate in the current science process and its non-sustainable operations, while also insisting for change in the direction toward future sustainable science operations, and hopefully with a rate of change such that this transformation occurs sooner rather than later due to the urgency of our circumstances. This is our first issue, and additional essential starting point issues, building on our five principles of holistic Life science, prior to planning scientific studies are as follows:

1. Orient science toward transitional goals that ultimately value Life. In anticipatory fashion, seek to embody and actualize Life value and to correct

the human−environment relation, so it becomes inherently win−win and environment can improve rather than degrade over time due to human actions and impacts.

2. Employ holism—seek more to unify, synthesize, and integrate rather than divide, differentiate, isolate, fragment, and analyze. We will always start with basics of systems and systems modeling—first steps always include defining the focal system, the system of study, the system boundaries, what is nonsystem (or environment, context, surroundings, etc.), and the variables of interest. This process—a conscious, intentional, and partly subjective process in which values are operating and mindsets, paradigms, assumptions (from our shared system of ideas) are employed—also involves deciding which variables will NOT be modeled and thus are treated as "exogenous" or "external drivers." We saw in Chapter 3 that we have a conscious choice whether Life and environment are separated versus integrated (see Fig. 3.2). And, we saw Patten's two ways to model environment—dualism versus synergism (Fig. 3.4).

3. Internalize all costs—bring science itself inside the system of study and plan strategically for science first to transition to sustainable operations and later to run using inherently sustainable infrastructure and principles. Running on renewable energy and recycling materials processes, and helping correct environmental quality metrics are required. Scientists have succeeded in developing many metrics of impact and monitoring progress and regress, such as genuine progress indicator (GPI, 2018), planetary boundaries (Rockström et al., 2009), and ecological footprint (Wackernagel et al., 2002); and it is time to turn these metrics inward on the practice of science itself. We are not so pedantic to claim that scientists should never use resources or travel (spreading research results, ideas, and face-to-face collaborations are part of the human experience and learning process), but we recognize that our own profession must change to operate in a context that is consistent with the whole goal of sustainability.

4. Keep Rosen's (1991) modeling relation in mind—see Fig. 6.1. This concise representation of the scientific process has many profound implications, and they are common sense despite having been largely ignored. One profound implication is that this is a *two-way relation* that operates and has action, impacts and real influence *in both directions*.

This starting point and anticipatory approach require additional explanation, so we take a brief detour to explain one key aspect here. (We examine in detail how Rosen's work helps bridge between science and the real world in Chapter 7.)

In the more commonly understood direction in Fig. 6.1, science makes models (formalisms, equations, relationships, theories, explanations, understanding, etc.) of Nature and how Nature works. But, Rosen says the process and "apparatus" of encoding (measurement) and decoding (predictions) exist partly in both realms. He also wrote that this modeling relation can operate in the opposite direction.

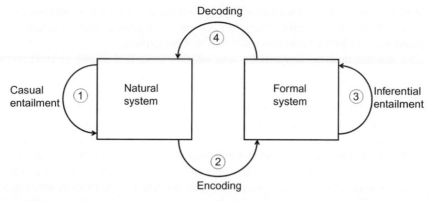

FIGURE 6.1

Modified slightly from Rosen's modeling relation. This appeared as his Figure 3H.2 on page 60 (Rosen, 1991). An elegant representation of the scientific process as heavily entangled with the real world it seeks to understand.

One can make a model of a natural system, or one can create a natural realization of a model (Rosen, 1991). Formal systems (ideas, mental models, theories, paradigms, etc.) can move from the mental and internal realm of the human self to become realized in physical and material form in the natural system or real world. This is tied to an extension and generalization of another observation of Rosen's. Writing about the most basic ideas of science and its validity, he wrote (Rosen, 1991):

> As philosophers have pointed out for millennia, all we perceive directly are ourselves, together with sensations and impressions that we normally interpret as coming from "outside" (i.e., from the ambience), and that we merely *impute*, as properties and predicates, to things in that ambience. The things themselves, the *noumena*, as Kant calls them, are inherently unknowable except through the perceptions they elicit in us; what we observe are *phenomena*, which are to an equally unknowable extent corrupted by our perceptual apparatus itself (which of course also sits partly in the ambience). (p. 56)

Thus, Rosen is saying that encoding and decoding in the modeling relation (see Fig. 6.1) and in the science process (as measurement and prediction, for example) exist in between the "inner" realm of formal systems, the mind and human self, and the "outer" realm of natural systems, the ambience, and the environment. And he is saying that our "perceptual apparatus" likewise exists straddling between these two realms—our eyes, ears, finger tips, and all sensory systems, as well as attendant motor systems that enable us to manipulate equipment or construct experiments. These are hybrid systems that bridge between and share fundamental aspects with the mental and abstract realm of the mind and the physical and material aspect of the environment that we both study, make, and are made of as we eat, breathe, excrete wastes, build, modify, use resources, and coexist. We are of the world around us. Life's great advance as a complex adaptive system was to emerge as differentiations from and using the stuff of the background and simultaneously develop ways to interact and make sense of this background.

We discuss in detail a synthetic idea on this topic tied to our main thesis of this book in Chapter 7. This is the idea that the machine metaphor, as the dominant mental model of the mainstream science paradigm, operates via the modeling relation (the two-way process of science) to alter the environment in detrimental ways. We now see Life and the environment breaking down and wearing out just like the machines we have imagined them to be. For now, we urge that we always keep in mind that science is an active, participatory, and two-way process that has multiple forms of impact on the systems of study and the world as a whole. By no means is it merely a passive or one-way process by which inert information moves from Nature into our minds, models, and shared bodies of knowledge.

Alexander (1964) added another layer to the model above to show explicitly the role of the mental model in interpreting the *noumena* (Fig. 6.2). Also, he put

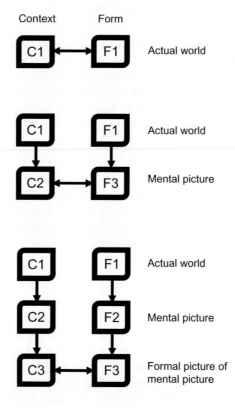

FIGURE 6.2

Mapping of context into form, according to level of abstraction. Redrawn from Alexander (1964).

Alexander, C., 1964. *Notes on the Synthesis of Form. Harvard University Press, Cambridge, MA, 216 pp.*

emphasis on the formal mathematical systems to highlight the added benefit that we can now apply logic, algebra, calculus, statistics, network analysis, simulation modeling, or any other number of techniques in which the "rules of the game" are known, repeatable, and result in "isomorphic" mappings. While it is unlikely if not impossible that two individuals will independently share an identical mental model and reach the same formal representation, once in its formal representation, a common language exists for further manipulation and investigation. Rosen (1991) also wrote about mappings or modeling relations between two formal systems.

The idea that the scientific process and enterprise (including all scientists and ourselves) are intimately entangled with the world it studies, seeks to understand, operates in, and seeks to develop technologies and applications for use in provides a fitting segue to our seven Life Lessons. We jump back in time to gain from insights of many others including Alfred Lotka. His holistic work we cite below was first published in 1925, but in our view, it has been sorely underappreciated for nearly 100 years.

Lesson 1. Coupled Complementary Life has Discrete and Sustained Aspects. Life's unique capacities for sustained existence and evolutionary improvement depend on coupled complementary processes and the functional unity of the discrete Life of organisms and the sustained Life of ecosystems and the biosphere.

In multiple sections of Chapters 1—3, we have discussed, described, and provided corroborating references for two distinct and complementary aspects of Life—the "discrete Life" most applicable to organisms and individuals (and even species) and the "sustained Life" most applicable to ecosystems and the biosphere. We also described how this dual-model view of Life is similar to complementary models of light as either/both particle and/or wave—in both cases, two disparate yet complementary models are needed for complete understanding, and the complex aspects of light and Life cannot be reduced to any single model.

These larger scales of Life organization—ecosystem as well as biosphere—are realms of study most often explored by those working in systems ecology, ecological networks, and global change—related fields. Studies at these levels of Life—environment organization are different from studies of the life of organisms and individuals, which can usually be determined as either alive or dead and where these are distinct states or conditions. Ecosystems and biosphere, on the other hand, always involve and depend on integration of elements and functions of both living and nonliving aspects. The discretely living forms have finite lifespans and necessary death; the sustained living forms, as far as we know, can continue to exist into the open-ended future. The interdependence of these two is based in part on the ways dead organisms are recycled via feeding relations, decomposition, and food web processes that feed Life's renewal. As we quoted Berry (1981) in Chapter 1 to say of the discrete Life forms "They die into each other's life, live into each other's death." We still have not found any other way to organize a biosphere, as Patten asked us to consider, other than by eating and being eaten. Thus, we propose to reflect on it, work hard to understand it, and make the best of this fundamental nature of Life.

Recognition of Life's distinct discrete and sustained aspects becomes powerful when we consider the radically empirical thought experiment we suggest reveals the need for a fundamental paradigm shift in sciences of Life and environment. The experimental question is what happens when one actually separates life from environment? We used this thought experiment in Chapter 5 to help illustrate the principle of radical empiricism, and it is useful again here. Unless we actually perform this experiment, either mentally or physically, we have no easily recognizable consequences for blurring the lines between, or failing to acknowledge the need for, the discrete and sustained Life categories. While not easily and immediately recognizable as consequences, we have proposed that the systemic symptoms of environmental degradation in Chapter 1 are indirect consequences of this conceptual choice (and myriad linked actions) to separate Life from environment, in a zero-sum, win−lose, finite game of interactions.

One can actually separate Life from environment—such as by placing an organism in a sealed container. This is where we see the benefit of the thought experiment relative to the real, physical experiment—there is no need to harm an organism to understand these concepts. If we actually separated an organism from its environment, then the organism would die. We propose a holistic interpretation of this experiment: it reveals that the organism died, that Life was destroyed because the discrete aspect of Life was isolated, disconnected, cut off from the sustained aspect of Life.

These experiments are reminiscent of ones done by Priestley (and others) leading up to his 1774 discovery that air is made of a composition of gases including oxygen (although he did not use that term). These early scientists subjected mice and other animals to sealed chambers in which they would either live or die depending on the experiment and its effect on the oxygen. They also used flames, which by comparison is better for causing no harm to living things.

If we modify the experiment and inside the sealed container place a more complete ecosystem as a realization of discrete Life unified with sustained Life—with plants, animals, and decomposers, for example—then the outcome would be different. This experiment was much like the Biosphere II project attempted in the early 1990s in the Arizona desert. But, one cannot avoid the question of scale, knowing what we do that many of Life's processes (oxygen formation, climate regulation, hydrologic cycle, etc.) are truly global. In the modified experiment, we have maintained the essential unity of discrete and sustained Life forms coexisting and operating together, and only in this more complete Life system are we able to see and realize both discrete and sustained aspects of Life. The individual organisms can only sustain their individual lives by actively participating in the context of the larger Life of the food web, community, and ecosystem as a functional whole.

An attempt at sustained Life at a smaller scale is given by the company, EcoSphere, who developed closed ecosystems that represent physical versions of this second experimental case—closed and sealed clear glass orbs containing multiple types of organisms in water. Their website states (www.eco-sphere.com):

Inside each EcoSphere are active micro-organisms, small shrimp, algae and bacteria, each existing in filtered sea water. Because the EcoSphere is a self-sustaining ecosystem, you never have to feed the life within. Simply provide your EcoSphere with a source of indirect natural or artificial light and enjoy this aesthetic blend of art and science, beauty and balance.

The website also describes how the closed ecosystems started from NASA research on whether "animal and plant life could be used to sustain humans in space exploration." The stakes now hit even closer to home—we imagine a human in a "sealed jar" instead of the mouse. Their history tab states:

Pioneered by Clair Folsom of the University of Washington research in the '60s and then jointly with Joe Hansen of NASA's Jet Propulsion Lab in the early '80s, along with other top bio-scientists, it was discovered that diverse colonies of microbes, [algae], and higher life forms (colonies) could persist apparently indefinitely in closed lab beakers.

They also tell how these lab beakers became a commercial product:

In 1983 Loren Acker, President of Engineering and Research Associates, Inc. (a small Arizona based medical research company) and long associate with the U.S. space program was visiting a NASA official and noticed one of the flasks Joe Hansen had created. Interested in the idea, Acker obtained a NASA Spin-Off Technology license for the EcoSphere. Along with a young employee, Daniel Harmony, they developed the first pilot manufacturing facility for EcoSpheres where the public and educators could directly benefit from NASA's works.

We quote at length this current website information for two reasons. First, this tells the fun and interesting story of research on space exploration as the context in which arose a radically empirical test of our dictionary and textbook definitions of Life as fundamentally organismal. Second, this story provides a link to consider again the Sustainer versus Transcender modes (Chapter 1), and the inherent complementarity of yet another pair of seeming opposites. These quotes indicate just how intimately entwined any effort to colonize space—the ultimate act of transcending Earth's limits—must be with efforts to sustain Life, both for astronauts and their essential Life support systems. It will not be enough to send colonizers out with supplies for the journey and beyond; they will need to sooner, rather than later, tap in to the Life sustaining properties already observed in full bloom on Earth. In fact, this shows a new twist that sustaining does not just mean maintaining but also creating and constructing.

Our original interest in a closely related concept—the idea of coupled complementary processes inherent in autotrophic and heterotrophic organism types—was assisted by the work of Lotka (1925) on "coupled transformers." Lotka wrote:

Coupled transformers are presented to us in profuse abundance, wherever one species feeds on another, so that the energy sink of one is the energy source of the other.

A compound transformer of this kind which is of very special interest is that composed of a plant species and an animal species feeding upon the former. The special virtue of this combination is as follows. The animal (catabiotic) species alone could not exist at all, since animals cannot anabolise inorganic food. The plant species alone, on the other hand, would have a very slow working cycle, because the decomposition of dead plant matter, and its reconstitution into CO_2, completing the cycle of its transformations, is very slow in the absence of animals, or at any rate very much slower than when the plant is consumed by animals and oxidized in their bodies. Thus the compound transformer (plant and animal) is very much more effective than the plant alone. (p. 330)

Lotka wrote that a plant could exist alone but with a "very slow working cycle", and that the coupled transformer of plant-plus-animal is "very much more effective." We appreciate his early and holistic recognition of this essential aspect of Life, but we would modify these last two statements. We do not think a plant or any autotroph could exist alone—such as isolated in a sealed container—as eventually some essential nutrient (CO_2 or inorganic nitrogen for example) would run out causing the plant to die before it could be broken down by abiotic processes. Thus, beyond merely being more effective, we see Lotka's "compound transformer" to represent a necessary holistic Life entity. These two complementary feeding aspects form the core functional components needed for the realization of sustained Life (autotrophic means self-feeding, heterotrophic means other-feeding). Lotka himself wrote of the need to see and treat Life forms holistically and that recognizing their integral unity is the best way to think about Life. He wrote (Lotka, 1925):

The several individual organisms of one species form in the aggregate one large transformer built up of many units operating in parallel.

... the entire body of all these species of organisms, together with certain inorganic structures, constitute one great world-wide transformer. It is well to accustom the mind to think of this as one vast unit, one great empire. (p. 330−331)

Here, Lotka not only points to the two types of organisms but also includes "certain inorganic structures" as integral to what he elsewhere called "the great world engine." By integrating the environment, Lotka has identified the essential Life−environment unity, and the irreducible importance of the biosphere as a Life unit-model and holon. This recognition fits with work of Vernadsky (1998 after 1926 original) who coined the term biosphere and Lovelock (Lovelock, 1972; Lovelock and Margulis, 1974) who conceived, and with Lynn Margulis modified, the Gaia hypothesis for how the Earth functions as a self-regulating, cybernetic system, which some have seen similar to a single living entity.

Lotka interestingly speculated on where we find plant (which he described as anabolic) and animal (catabolic) functional forms in present-day ecosystems. He wrote:

> It is, of course, conceivable that the anabolic and catabolic functions should, in their entirety of a complete cycle, be combined in one structure, one organism. Physically, there is no reason why this should not be, and, in fact, nature has made some abortive attempts to develop the plant-animal type of organism; there are a limited number of plants that assimilate animal food, and there are a few animals, such as *Hydra viridis*, that assimilate carbon dioxide from the air by the aid of chlorophyll. But these are exceptions, freaks of nature, so to speak. For some reason these mixed types have not gained for themselves a significant position in the scheme of nature. Selection, evolution, has altogether favored the compound type of transformer, splitting the anabolic and the catabolic functions, and assigning the major share of each to a separate organism. (p. 330)

Rather than a process of *splitting* two anabolic and catabolic functions, we see Lotka's observations to reveal that evolution, selection, and the original and fundamental nature of Life to have *unified* three functions and realms—the anabolic, catabolic, and ambient environment—much as in the unified "worldwide transformer" as "one vast unit" he described above.

We also note that novel methods such as environmental genomics may provide new insights into this. For example, Chivian et al. (2008) reported the discovery of a "single-species ecosystem" deep within the Earth.

In Chapter 4, we cited the origin of Life scenario of Odum (1971), which he described as "ecological system precedes the origin of life." Instead of anabolic and catabolic, or plant and animal, Odum wrote of production and consumption (or respiration) and "adding molecules" and "segmentation." This scenario included an essential environmental function provided by "circulating seas" that provided a closed cycle and worked to move products of production (composed polymers, anabolic process) to deeper regions in a water body where they could be consumed (decomposed polymers, catabolic process) and later return products of consumption back toward the surface and "photochemical zone" for the next round of production. Adding another variation on the terms for identification of these various complementary functions in Life, Fiscus (2001−2002) referred to them as "composers" and "decomposers," each like the opposite of the other and again operating in coupled complementary and coupled transformer fashion.

In Chapter 5, in the Life science principle of holism, we saw that Bateson identified the unified system of life−environment, "organism plus environment" (or "organism-in-its-environment") as he referred to it, as the best idea for a "unit of survival under natural selection." Odum (1971) saw the complete cycle including two Life functions and the environment as a "choice-loop-selector," his version of a Life—environment model for the units of Life and evolution. He noted that this is the essential functional configuration of the present-day biosphere, as well as crucial to understanding the original and fundamental nature of Life. He saw the photochemical reaction in his origin of Life system as a "random-choice generator" and the abiotic hydrologic circulation cycle to provide "a reward loop to choose for further use those units capable of the stages of evolution" at appropriate areas in cycle. These functions all collaborated to improve the system as a whole: "Increased order is paid for energetically in small increments."

Lotka (1925) wrote of a similar view as integrated with his scientific strategy:

> It is customary to discuss the "evolution of a species of organisms." As we proceed we shall see many reasons why we should constantly take in view the evolution, as a whole, of the system (organism plus environment). ... the physical laws governing evolution in all probability take on a simpler form when referred to the system as a whole than to any portion thereof. (p. 16, footnote 19)

And later, with more emphasis on the benefit of holistic strategy for finding fundamentals:

> Biologists have rather been in the habit of reflecting upon the evolution of individual species. This point of view does not bear the promise of success, if our aim is to find expression for the fundamental law of evolution. We shall probably fare better if we constantly recall that the physical object before us is an undivided system, that the divisions we make therein are more or less arbitrary importations, psychological rather than physical, and as such, are likely to introduce complications into the expression of natural laws operating upon the system as a whole. (p. 158)

And later:

> ... the concept of evolution, to serve us in its full utility, must be applied, not to an individual species, but to groups of species which evolve in mutual interdependence; and further to the system as a whole, of which such groups form inseparable part. (p. 277)

An interesting and ironic historical note is that Lotka is perhaps most famous in ecology and related disciplines for his model of predator—prey dynamics, usually referred to as the Lotka—Volterra model. Yet, this one useful contribution of Lotka's was extracted by taking something out of context, one of the dangers of unbalanced reductionism and analysis of which we have warned. Had Lotka's full theoretical and scientific approach been understood in 1925, we might have no need for our book at this time, and we might have avoided our current suite of Life-threatening environmental dysfunction symptoms.

From the two essential complementary functions as integrated with the environment, we see Life to achieve emergent holistic capacities that are beyond a mere sum of these parts. This includes the unique capacity for both discrete and sustained Life, with each aiding the other, and the capacity for Life to improve itself and its environment as it operates.

In Chapter 8, we will see models for technologies using "differential systems" that provide special capacities much like the synergy from coupled transformers and coupled complementary processes here.

While we have just looked at complementarity of two collaborative processes, we next learn of 10 system-wide tendencies that are complementary in the view they afford of complex systems.

Lesson 2. Complementarity of Ecological Goal Functions. Ecosystems display a remarkable tenacity to grow and develop under whatever prevailing conditions are present in terms of solar radiation, temperature, precipitation, nutrients availability, and species diversity. This movement away from simplicity to greater levels of organization, complexity, and diversity is what identifies them as far from equilibrium systems. Tracking key variables that change over time during this growth and development has interested ecologists during the past century. Notably, systems ecologists have hypothesized that ecosystems seek to maximize biomass storage and to maximize energy dissipation, and these two seemingly contradictory trends can both increase when residence time is also increasing and specific dissipation decreasing. The mutual consistency of these and other seemingly disparate ecological tendencies can be explained due to network organization.

Fath et al. (2001) worked to compare, contrast, and interrelate a set of 10 ecological whole-system tendencies that had been proposed to explain how ecosystems grow, develop, and change over time. They also referred to these principles as "goal functions," "extremal principles," "orientors," and "attractors," and they considered whether ecosystem networks change along trajectories in ways to maximize (steadily increase) or minimize (steadily decrease) a given goal function over time. The principles they studied, and the motivating idea to characterize trends in ecosystem development, were often inspired or informed by the landmark paper of Odum (1969) in which he hypothesized a suite of such trends. As with the whole-system Life Lessons above, these principles fully require comprehensive data from ecosystem networks rather than narrowly restricted studies of small subsets of communities or ecosystems.

Another motivation for this synthesis study was their recognition that two classical principles from thermodynamics do not apply to ecosystems. The second law of thermodynamics does not apply directly, as ecosystems are not isolated systems. The principle of decreasing entropy production for open dissipative systems near steady state (Nicolis and Prigogine, 1977) does not apply, as ecosystems operate under conditions far from equilibrium. While many of the goal functions they studied apply to material flows, their primary approach was to consider energy flow and storage in networks of living systems.

The 10 organizing principles they studied are the tendency of ecosystem networks to (1) maximize power, (2) maximize storage, (3) maximize empower, (4) maximize emergy, (5) maximize ascendency, (6) maximize dissipation, (7) maximize cycling, (8) maximize residence time, (9) minimize specific dissipation, and (10) minimize empower to exergy ratio. These system properties were developed and proposed by a variety of ecologists, biologists, and systems thinkers including Alfred Lotka, Howard Odum, Robert Ulanowicz, Sven Jørgensen, Harold Morowitz, and others. After explaining and examining the mathematical formulations for calculating these whole-system measures, they concluded (Fath et al., 2001):

> Much debate and confusion have centered on the appropriateness of these various goal functions because, at first glance, the simultaneous realization of [the ten goal functions] seems contradictory. Further inspection, however, shows that all these goal functions are in fact mutually consistent. They are all generated by network processes and they give complementary perspectives on the spontaneous directions of ecological growth and development. (p. 502)

While Fath et al. (2001) studied the 10 goal functions using network environ theory, they also noted that only those principles linked to empower and emergy (terms and measures created by Howard Odum to signify embodied power and embodied energy) and to ascendency of Ulanowicz were derived explicitly from network studies. The fact that many of the goal functions came from other methods of systems analysis yet yielded complementary results led to another realization from their article, again describing the variety of goal functions and their sources:

> Only those pertaining to empower, emergy and ascendency were originally conceived in an explicit network context, yet it is global systemic organization that is behind the similarities inherent in all the studied goal functions. The implication is that the network perspective is fundamental, and somehow the originators of the different orientors managed to capture this intuitively in their concepts. (p. 504)

They also described the interplay of these whole-system functions using this maxim of the organizing principles of how ecosystems operate:

> Get as much as you can (maximize input and first-passage flow), hold on to it for as long as you can (maximize retention time), and if you must let it go, then try to get it back (maximize cycling).

They concluded that minimization of specific dissipation is the "most encompassing" principle as it captures all three of the principles in the maxim above while also reconciling how maximum storage and maximum dissipation can occur simultaneously while their ratio is minimizing.

The need for multiple whole-system principles to represent the behavior of ecosystem networks, and the Fath et al. (2001) approach to find their mutually consistency, fits with the need for our principle of complexity in holistic Life science (Chapter 5). These network science goal functions also help to make sense of the win—win environmental outcomes we noted in Chapter 3 as beneficial outcomes of the Life—environment relation. For example, both the formation and maintenance of the oxygen atmosphere, and the formation and maintenance of soils, reflect the complex interplay of both increasing storage (in the buildup of stocks) and increasing dissipation (via input, use, metabolic processing, or environmental flux due to erosion), while network cycling and retention time serve to unify the two seemingly opposing trends.

The next two Life Lessons come from studies that require comprehensive network datasets such as food webs. Once above a certain threshold of complexity, number of interdependent nodes, or wholeness of description, we see quantitative evidence of the surprising reality of Life in its ecosystem and network form. As with all of the seven Lessons, these next two results apply to human socio-economic networks as well.

Lesson 3. Dominance of Indirect Effects. Indirect relations and impacts between Life forms are usually greater than direct relations and impacts and are often of different quality.

Lesson 1, that Life depends on coupled complementary processes, and Lesson 4, that all living things impact each other in measurable transfers of energy and matter, both indicate results that cannot be obtained from studies more narrowly focused, on heavily isolated subsets, than a whole ecosystem food web. Said another way, this lesson from Life science and ENA requires that holistic, comprehensive studies of whole networks/systems are essential to understanding Life. We do not suggest that every study must be so holistic and comprehensive; but in the sense of balance and synergy between analysis and synthesis, and between reductionism and holism, perhaps one-half of all studies should be comparably complete in terms of measurements taken on all participants in a given community, ecosystem, or Life system of study. The level of aggregation is an ongoing area of research, but at a minimum including the most basic aspects of an ecosystem would have around six functional groups (Fath, 2004).

Life Lesson 3 is similar in this regard—the next fundamental observation and discovery from ENA and Patten's counterpart methods, network environ analysis (NEA)—cannot be seen from isolated analysis alone; and thus, requires holistic studies of complete (or as nearly complete as practically possible) communities, ecosystems, networks, food webs, and Life systems.

This quote from Patten (2016a) helps to describe both Lessons 3 and 4 and to explain their paradigm-shifting significance:

> Ecological systems at all scales are, by their higher order indirect pathways, more highly inter-connected and interdependent than denoted by adjacent, first order linkages alone. In ecosystems, in particular: the pathways of food chains hidden in food webs are much longer, energy cycling is the rule, and food web constituents are more interrelated than usually described in empirical studies. All these properties run counter to currently accepted ecological thought. (p. 64—65)

Patten (2016a) referred to Lesson 3 as "network nonlocality," while we are describing this as "dominance of indirect effects." He wrote:

> Network nonlocality underscores the essential holism of all reasonably well-connected transactional systems. Every component in ecosystems, large and sparse in adjacent linkages though these may be, are richly interconnected to a multitude of others by interactions at a network distance, only few of which have direct, or even adjacent links to the one in question. Indirect effects are the glue of holistic organization. ... (p. 65)

The paper that Patten notes as the key reference (Higashi and Patten, 1989) describes the first study in which the dominance of indirect effects was clearly established. This article built on previous work of Patten (1982) on a network model of the continental shelf ecosystem in the Gulf of Mexico with nine nodes (compartments) (Fig. 6.3). This model tracked flows of carbon between aggregated functional groups such as pelagic planktivores, pelagic omnivores, plankton, benthos, etc.

Higashi and Patten (1989) describe the mathematical and scientific steps they took to define and quantify both direct and indirect effects. Their methods are similar to Ulanowicz and Puccia (1990) above, although with added consideration of temporal factors related to time lags associated with any transfer of energy or material between network nodes (food web participants).

In the 9×9 cell matrix of possible interactions of the nine-node network, Higashi and Patten reported the following:

1. 21 had no direct or indirect interactions. These are associated with X3 and X7, which have no outflow, thus only direct interactions into them are present. And, X8, Plankton, which has no inflow from within the system.
2. 33 showed positive indirect effects despite no direct effects.
3. 27 have direct interactions, of which 8 (30%) have greater indirect influence than the direct flow.

They also reported that the total of indirect effects considering all compartments was 3.1 times (and thus much greater than) the total of all direct effects. This result, that indirect effects are greater than direct ones, has been corroborated in models of numerous ecosystem networks (Salas and Borrett, 2011).

Patten (2016a) later credited his primary "cardinal hypothesis" of Holoecology, network proliferation, as the key reason behind this principle of the dominance of indirect effects. He wrote that his hypothesis of network proliferation "concerns the exponential increase in pathway numbers with length between each pair of nodes (compartments) in an interconnected system."

There are research articles that explain the technical aspects, but some of the basics are helpful to get a feel for the approach used here. The first step is to recognize that any network can be represented by a matrix where the values in the matrix correspond to whether or not there is a connection in the network. An adjacency matrix is the term used for the presence or absence of connections

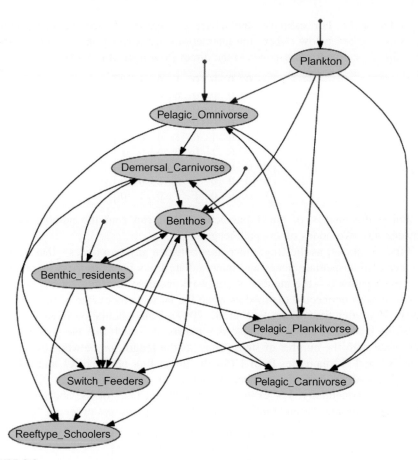

FIGURE 6.3

This is a typical ecological network showing the flow of energy between compartments. Each circle represents an ecological functional group (or species) and the arrows represent the flow of energy between compartments. X1 = pelagic planktivores, X2 = pelagic omnivores, X3 = pelagic carnivores, X4 = demersal carnivores, X5 = switch feeders, X6 = benthic residents, X7 = reeftype schoolers, X8 = plankton, X9 = benthos (after Patten, 1982. Environs: relativistic elementary particles for ecology. American Naturalist. 119(2), 179–219). This figure was constructed using the online ENA tool EcoNet http://eco.engr.uga.edu/cgi-bin/econetV2.cgi. Other software tools are available to perform the network analysis routines described below such as the MATLAB function nea. m (www.mathworks.com/matlabcentral/fileexchange/5261-nea-m) and enaR (https://cran. r-project.org/web/packages/enaR/index.html).

using 0s or 1s. For example, the adjacency matrix, A, for the ecosystem in Fig. 6.3 is given below (where the interactions are given from columns to rows; thus, the a_{21} element corresponds to the direct flow from $X1 \rightarrow X2$):

$$A = \begin{bmatrix} 0 & 0 & 0 & 0 & 0 & 1 & 0 & 1 & 0 \\ 1 & 0 & 0 & 0 & 0 & 0 & 0 & 1 & 0 \\ 1 & 1 & 0 & 0 & 0 & 1 & 0 & 1 & 1 \\ 1 & 1 & 0 & 0 & 0 & 1 & 0 & 0 & 0 \\ 1 & 1 & 0 & 1 & 0 & 1 & 0 & 0 & 1 \\ 0 & 0 & 0 & 1 & 0 & 0 & 0 & 0 & 1 \\ 0 & 0 & 0 & 1 & 1 & 1 & 0 & 0 & 1 \\ 0 & 0 & 0 & 0 & 0 & 0 & 0 & 0 & 0 \\ 1 & 0 & 0 & 0 & 1 & 1 & 0 & 1 & 0 \end{bmatrix}$$

Using this matrix of direct interactions, one can find the number of paths between any two nodes of any path length.

First, as above, as the length increases between any two nodes—that is, as the number of intermediate nodes and intermediate links between them increases for any given pathway—the number of pathways increases combinatorially when the system is well connected (defined as when the largest eigenvalue of the adjacency matrix is greater than 1, see Fath et al., 2007). This calculation is easy to make by taking powers of the direct adjacency matrix, A^m, such that the power is commensurate with the number of steps. Thus, as the length, m, between two components X1 and X2 increases from 1 (for the direct link between X1 and X2) to 2, 3, 5, 10, and beyond, the number of pathways connecting nodes X1 and X2 increases exponentially. For example, in the Patten Gulf of Mexico model cited above, Higashi and Patten (1989) reported 134,280 pathways of length 10.

$$A^2 = \begin{bmatrix} 1 & 0 & 0 & 1 & 0 & 2 & 0 & 2 & 1 \\ 2 & 1 & 0 & 0 & 0 & 1 & 0 & 3 & 0 \\ 4 & 2 & 1 & 1 & 1 & 4 & 0 & 5 & 3 \\ 3 & 2 & 0 & 2 & 0 & 3 & 0 & 2 & 1 \\ 5 & 3 & 0 & 3 & 2 & 5 & 0 & 3 & 3 \\ 2 & 1 & 0 & 2 & 1 & 3 & 0 & 1 & 2 \\ 3 & 2 & 0 & 4 & 3 & 5 & 1 & 1 & 4 \\ 0 & 0 & 0 & 0 & 0 & 0 & 0 & 1 & 0 \\ 3 & 1 & 0 & 2 & 2 & 4 & 0 & 3 & 3 \end{bmatrix}$$

And, the number of pathways that are of length 10 is given by A^{10}, where:

$$A^{10} = \begin{bmatrix} 13,263 & 6193 & 0 & 9397 & 4260 & 16,467 & 0 & 13,083 & 9397 \\ 6014 & 2810 & 0 & 4260 & 1933 & 7464 & 0 & 5927 & 4260 \\ 41,273 & 19,276 & 1 & 29,238 & 13,271 & 51,236 & 0 & 40,696 & 29,248 \\ 21,998 & 10,274 & 0 & 15,591 & 7069 & 27,315 & 0 & 21,696 & 15,590 \\ 51,245 & 23,931 & 0 & 36,317 & 16,467 & 63,631 & 0 & 50,548 & 36,317 \\ 29,247 & 13,657 & 0 & 20,727 & 9397 & 36,317 & 0 & 28,852 & 20,727 \\ 65,743 & 30,698 & 0 & 46,590 & 21,122 & 81,635 & 1 & 64,860 & 46,590 \\ 0 & 0 & 0 & 0 & 0 & 0 & 0 & 1 & 0 \\ 42,510 & 19,850 & 0 & 30,123 & 13,658 & 52,783 & 0 & 41,935 & 30,124 \end{bmatrix}$$

This shows clearly the rapid increase in the number of pathways over still relatively short lengths. Note, though, that the third and seventh columns are 0 except the diagonal element. This is because both compartments, Pelagic_Carnivores and Reeftype_Schoolers, are "sink" compartments with no outflow. There can be no indirect pathways emanating from these compartments. In Markov chain terminology, such nodes are referred to as absorbing states. Also, the eighth row is 0 except the diagonal, because there is no flow into the compartment from the other compartments in the network. This compartment, Plankton, only receives energy from sunlight which is outside the system boundary. Such a node is called a taboo state because no flows enter it. In spite of these three exceptions, there are very many pathways between the other nodes as the path length increases. The a_{ij} elements in the other cells will all go to infinity as the path length goes to infinity.

Second, as the length of pathway increases, those "concatenated" pathways carry less and less substance (energy or material, such as carbon, nitrogen, etc.).

Finally, since there are very many more of the longer pathways, even though the longer pathways individually carry less substance, cumulatively these indirect pathways carry more aggregate substance than direct pathways. Using marker molecules, one could trace the reticulated fabric that weaves through all Life, in forests, oceans, farms, soils, cities, and any other systems.

Patten (2016a) concludes this principle stating that except for "very small, minimally connected models," higher order indirect effects are greater than first-order direct effects in ecosystems. As we said above, given that all Life forms must exist and participate in ecosystems and food webs, this result thus becomes generic and applicable for all Life.

We mention only one other extension to this very profound result, our Life Lesson 3. Patten (1981) wrote of the implications for evolutionary theory and importance of his environ theory as a means to characterize ecological niches:

Evolution proceeds by natural selection of heritable variations of individual organisms based on direct influences of environment. However, indirect effects probably vastly outweigh direct ones in ecosystems. Therefore, why is evolution based on direct effects only? The ecological niche represents the point of direct contact between organisms and their environments. To encompass indirect influences, niches are extended to new structures, environs, which are units of organism-environment coevolution. The motive force for coevolution is closure of outputs back upon inputs of the organism members of ecosystems. Closure is achieved by biogeochemical cycling and feedback interactions, direct and indirect, between organisms. To the extent that closure does not occur, there is no imperative for organism-environment coevolution. Coevolution at the system level based on indirect effects is compatible with normal evolution at the individual organism level based on direct effects. The organism is the unit of the latter, but environs are the unit of coevolution. (p. 845)

And a bit later:

> Thus, a paradox arises. If indirect influences are the most important, why is evolution based on direct effects only? (p. 845)

Thus, the dominance of indirect effects calls into question the fundamental ideas we have come to take for granted with the theory of evolution by natural selection. Note also that here Patten touches on other crucial holistic concepts we have discussed, including system closure and two essential hierarchical scales or holons—organisms and ecosystems.

We next consider another fundamental result obtained using both ENA and NEA.

Lesson 4. All Life is Physically and Relationally Connected. Via ENA we can quantify how the connections, relations, flows of energy and material between life forms are crucial to understanding Life. These network relations between all entities are equally as important as more traditional approaches focusing on dynamics (as in dynamic process models) and measures of materials such as the numbers, sizes, movements, or masses of individual entities.

ENA has shown the quantifiable, tangible, physical, and material hard evidence to back a widely stated concept of wholeness and connection. The notion of environmental unity has been expressed by scientists, naturalists, and poets, among many others. Isaac Asimov expressed it:

> What makes it so hard to organize the environment sensibly is that everything we touch is hooked up to everything else. (Asimov and Shulman, 1988)

As did John Muir before:

> When we try to pick out anything by itself, we find it hitched to everything else in the Universe. (Muir, 1911).

This intuitive notion relative to environment and universe extends to human interdependence in John Donne's meditation on death (Donne, 1997):

> No man is an island, entire of itself; every man is a piece of the continent, a part of the main. If a clod be washed away by the sea, Europe is the less, as well as if a promontory were, as well as if a manor of thy friend's or of thine own were: any man's death diminishes me, because I am involved in mankind, and therefore never send to know for whom the bells tolls; it tolls for thee. (p. 120)

The scientific methods of ENA, operating with data of food webs to produce the calculated matrix of total impacts (Ulanowicz and Puccia, 1990; Patten, 1991; Ulanowicz, 2004), transform these aphorisms into permanent scientific principles and, as far as we are aware, universally observed aspects of the reality of interconnectedness of Life. The extension from feeding relations to all of Life is valid, as all living things eat and are eaten, even if for some life forms (such as plants and autotrophic organisms), most of their food comes from nonliving environmental resources (such as sun light, CO_2, water).

An important calculated product on the way to holistic network analysis—the matrix of total impacts—demonstrates that "everything is connected to everything" very literally and quantifiably. This is demonstrated by a matrix filled with nonzero values registering measurable impacts between any two organisms or functional groups in an ecosystem food web. Thus, this idea of radical connection and interdependence is not just a loose metaphor, romantic notion, or philosophical speculation. This holism is a bona fide, measurable, and ubiquitous fact.

Ulanowicz and Puccia (1990) stated this fact with exclamation and via the ideas of pathways and connections (note that a "graph" is another term for network of nodes and vertices, and a weighted digraph is a quantified food web network)—"One or more pathways will then connect any two elements in the graph, so that every node is literally connected with every other in the system!" However, they also noted that "not all connections are equivalent, and it remains somehow to ordinate the indirect impacts...." We provide a brief narrative summary of the method they developed to ordinate or quantify the total impacts including all direct and indirect impacts between all nodes in a food web network. For the mathematical, graphical, and quantitative descriptions, see Ulanowicz and Puccia (1990) or Ulanowicz (2004).

They started by observing the need to quantify two forms of impact that occur every time a predator feeds on prey—the positive impact for the predator, and the negative impact for the prey. We use this common example of predator and prey, but the logic, methods, and results apply to all ecosystem transfers of conservative substances such as energy and material, with modifications as needed and as described below. They quantified the positive impact for the predator by normalizing the energy (or nitrogen, carbon, etc. depending on the currencies of the food web network) gained from the individual prey item in question as compared with energy gained from all the prey items of the given predator. To normalize the positive benefit in an item of prey eaten is to set it relative to (divide it by) the total intake of energy or other sustenance from all items in the diet.

The second step they developed was for weighting the negative impact of predator on prey. They again normalized or measured this in a relative sense based on the fraction of a prey species' total production (output from the prey compartment, increase in biomass, etc.) that is consumed by a given predator. They modified this calculation slightly by excluding respiratory output from the estimate of the total production (the denominator used for normalization and calculating the fraction). Thus, the denominator is the *net output* from the prey after subtracting respiratory costs or losses. One other calculation detail they used is for cases when the receiving compartment is not actually a predator but a passive compartment like detritus. For these transfers, the negative impact is set to 0 as "detrital flows usually do not directly impact their donors to the same negative degree as do active predators" (Ulanowicz and Puccia, 1990).

The final step for pairwise and direct interactions is to subtract the negative impacts from the positive impacts. Since the positive and negative impacts are all normalized, they range between 0 and 1. And after subtracting the negative from positive impacts, the net impacts range from -1 to $+1$. Each species or compartment in the network has a net impact value that indicates the net impacts between all pairs of species or nodes in the network. The resulting $n \times n$ matrix is the *net impact matrix.*

Next, Ulanowicz and Puccia (1990) use the matrix multiplication approach described above to quantify higher order interactions—the net impacts matrix is filled with first order, direct impacts between each pair of nodes. To calculate all direct and all indirect net impacts, they note that "the overall trophic impact of any concatenation of direct effects is measured by the product of all the [net impacts] along that pathway." The normalization scheme used for net impacts ensures that the infinite series converges to a finite limit, a familiar mathematical result in input—output theory by which total indirect effects are calculated. The infinite series of matrix multiplications produces the total net impacts between each pair of network participants over all direct and indirect pathways in the network.

While the resulting total net impacts (or total impacts, for brevity) between all participants, species, or nodes in the network may be very small in magnitude, and for some additional calculations or ecological interpretations, they are sometimes ignored when very small (e.g., $<1/10,000$), they are nonetheless all nonzero. These are the methods and results that prove that everything is connected to everything in the case of ecosystem food webs, which are universal integral aspects of the context of all Life.

The mathematics shows that the influence of one part of the system on another is a radical form of relational interdependence. While the physical flows establish the network, the relations—both direct and indirect—are informational and not conserved. In fact, it is not uncommon to find examples where the direct and the indirect relations are not the same. This quite clearly means that what you see is not always what you get. For example, a biologist making observations and taking notes in the field might record an act of predation such as an individual alligator consuming an individual frog. At the scale of the ecosystem food web, the alligator population can be represented by one box/compartment and the frog by another, and there would be an arrow indicating the flow of energy from the frog compartment to the alligators. A reductionistic analysis would stop, content in the knowledge that a predation relation exists between the two, ignoring the indirect effects of the entire food web network. The network analysis could show that the alligator population and the frog population are in fact mutualists, both benefitting from the presence of the other. Of course, this is no comfort to the individual frog that ended up in the mouth of the alligator, but it demonstrates a higher order of organization at the (eco)system level (and, also the important distinction between Life and life and Death and death). This is precisely the type of indirect relations that Bondavalli and Ulanowicz (1999) found in a comprehensive study of the Florida everglades. Overall, the charismatic, iconic, top predator alligator, in turn, was mutualistic with 11 of its prey items. Clearly, to get an accurate picture of the role of each population in the food web, and therefore, to conduct proper management of the ecosystem, it is necessary to consider the entire web of indirect interactions.

Returning to the network example from earlier, we can see these interdependencies emerge as all elements of the integral (summed over all direct and indirect pathways) relational interaction are nonzero. Starting first we construct the direct relational matrix which shows all the pairwise, zero-sum interactions—for every plus sign there is a corresponding negative sign (read across the main matrix diagonal). The notation that has been adopted for this is the D matrix, for direct relations, given below (note for scale the values are multiplied by 100):

$$D \times 100 = \begin{bmatrix} 0 & -2.9000 & -1.2 & -0.2000 & -0.5 & 7.0637 & 0 & 92.9363 & -0.9 \\ 0.1236 & 0 & -2.1 & -0.1000 & -0.4 & 0 & 0 & 0.8909 & 0 \\ 1.7935 & 73.6704 & 0 & 0 & 0 & 10.5572 & 0 & 10.4176 & 3.5613 \\ 1.0510 & 12.3349 & 0 & 0 & -1.2 & 85.7141 & -0.4 & 0 & 0 \\ 0.0213 & 0.4003 & 0 & 0.0097 & 0 & 0.3012 & -0.3 & 0 & -0.8968 \\ -0.3 & 0 & -0.3 & -0.6927 & -0.3 & 0 & -0.2 & 0 & 0.7156 \\ 0 & 0 & 0 & 0.4589 & 42.4193 & 28.3905 & 0 & 0 & 28.7313 \\ -4.0 & -0.9 & -0.3 & 0 & 0 & 0 & 0 & 0 & -0.8 \\ 0.0378 & 0 & -0.1 & 0 & 0.8827 & -0.7071 & -0.2 & 0.7801 & 0 \end{bmatrix}$$

Although the network is generally well connected, there are still some compartments that do not interact directly. However, when we take the power series of this matrix, giving the relations over indirect pathways, we see clearly that indeed all compartments are connected and related to all others, even if the strengths of the interactions differ. This is given in what is called the integral utility matrix, U, which sums utilities over all path lengths, below:

$$U \times 100 = \begin{bmatrix} 96.373 & -4.621 & -1.345 & -0.233 & -0.497 & 6.473 & -0.007 & 89.372 & -1.582 \\ 0.056 & 98.45 & -2.069 & -0.097 & -0.392 & -0.297 & 0.002 & 0.713 & -0.078 \\ 1.336 & 72.32 & 98.394 & -0.147 & -0.306 & 10.324 & -0.0216 & 12.164 & 3.464 \\ 0.764 & 11.844 & -0.516 & 99.395 & -1.734 & 85.027 & -0.563 & 0.765 & 0.431 \\ 0.020 & 0.394 & -0.008 & 0.006 & 99.862 & 0.229 & -0.298 & 0.014 & -0.980 \\ -0.298 & -0.287 & -0.288 & -0.688 & -0.364 & 99.298 & -0.196 & -0.305 & 0.652 \\ -0.071 & 0.118 & -0.116 & 0.265 & 42.479 & 28.458 & 99.758 & 0.146 & 28.49 \\ -3.86 & -0.918 & -0.222 & 0.011 & 0.018 & -0.281 & 0.002 & 96.38 & -0.746 \\ 0.0074 & -0.076 & -0.098 & 0.005 & 0.799 & -0.767 & -0.201 & 0.775 & 99.92 \end{bmatrix}$$

The principle that all parts of a system are connected through transactions and/or relations, and the ability to quantify those influences, has profound implications. This is powerful knowledge that should be of use in a science that attempts to avoid unwanted and unanticipated consequences. Of course, interaction strengths diminish with time and distance, but there are real connections that can have real influence on the outcomes.

Very small forces or influences can at times have great impact. For example, "the butterfly effect" is a related idea from chaos theory. The butterfly effect (Lorenz, 1963) says that the impact of a butterfly flapping its wings in one location on Earth can spawn a hurricane on the other side of the globe. This vignette is used to emphasize "sensitive dependence on initial conditions" inherent in systems exhibiting chaos. It also plausibly requires something like our Life Lesson 4 here—perhaps an atmospheric version of "everything is connected" that transmits and carries the butterfly wing flap influence across the globe to start the hurricane. This also suggests that despite the smaller interaction strengths after 10, 100, or 1000 food web links, if those interactions impact a chaotic system or one exhibiting self-organized criticality, the tiny influence could yield dramatic effect. The ecological vignette might be that a butterfly sipping nectar in one location can later feed a blue whale or elephant on the other side of the globe.

Lesson 5. Mutualism is Common and Crucial. "Life is beneficent under network rule."

One of the key findings that emerged from the network relational analysis described in the previous Life Lesson is that not only are all parts of the system connected but also that often the most dominate relation type that occurs is mutualism. Keeping in mind that the analysis was developed for ecological food webs, this seemingly flies in the face of the conventional wisdom that nature is dominated by competitive relations and is best described as "red in tooth and claw." Clearly, food webs are full of direct predator—prey feeding transactions, but behind those direct interactions are series and series of complex, indirect pathways and interactions that overall tend to give the network a mutualistic outlook. Many of those initial non-interactions in the direct matrix fill in as positive relations for both compartments. In other words, they both benefit each other by being part of the same network. One reason is because the transactions keep the circulation going such that the resources are more evenly and adequately shared by all the nodes. It is only when we isolate two interacting nodes, such as a predator eating a prey, out of the complex food web, that we see only the winners and losers (as we saw in Ulanowicz' reduced food web, Fig. 3.3). But, as we said in Lesson 1, just as one organism cannot be isolated from its environmental context, neither can only two be fragmented out. It's the interplay of the entire system that keeps the system going. There is positive feedback and autocatalytic cycling that spread the energy and benefits to all parts of the network.

Looking back at our concrete example of the 9×9 aquatic food web, we can look at the relation types by focusing on the signs of the elements in the U matrix, given by

$$sign(U) = \begin{bmatrix} + & - & - & - & - & + & - & + & - \\ + & + & - & - & - & - & + & + & - \\ + & + & + & - & - & + & - & + & + \\ + & + & - & + & - & + & - & + & + \\ + & + & - & + & + & + & - & + & - \\ - & - & - & - & - & + & - & - & + \\ - & + & - & + & + & + & + & + & + \\ - & - & - & + & + & - & + & + & - \\ + & - & - & + & + & - & - & + & + \end{bmatrix}$$

In this network model, of the 81 possible signs, 43 are positive and 38 are negative. More specifically, there are 19 mutualistic relations (including the self-mutualisms on the diagonal), 14 competition relations, and 24 pairs of consumer-resource "predations." In this case, the mutualisms arose from the null direct relations from five pairs:

Pelagic_omnivores and Reeftype_schoolers
Demersal_carnivores and Plankton
Switch_feeders and Plankton
Reeftype_Schoolers and Plankton
Demarsal_carnivores and Benthos

Direct null relations that switched to competition include seven pairs:

Pelagic_carnivores and Demersal_carnivores
Pelagic_carnivores and Switch_feeders
Benthic_residents and Reeftype_schoolers
Pelagic_omnivores and Benthic_residents
Switch_feeders and Pelagic_planktivores
Switch_feeders and Pelagic_canivores
Pelagic_omnivores and Benthos

Here, all the "new" relations emerged from the cases that were not directly interacting. However, there has been evidence of the relationship type switching from one type to another more drastically. For example, in some networks, the consumer-resource $(+, -)$ becomes mutualistic $(+, +)$. Most dramatically, we have seen examples where the direct $(+, -)$ actually reverses to become $(-, +)$. In other words, what we observe in the field, flow of energy (or nutrients or water) from one compartment to another, actually results in, when considering the network of indirect interactions, a benefit to the compartment being eaten and a cost to the one doing the eating (at the Life/Death scale, not the life/death scale).

There are numerous examples of food webs that display "beneficial predators" as common in ecosystem food webs. Ulanowicz and Puccia (1990) reported on the beneficial impacts of predators in a 36-compartment network for the Chesapeake mesohaline ecosystem. They found six instances of beneficial predators, which include spot feeding on crustacean deposit feeders, ctenophores feeding on bacteria, bay anchovy feeding in phytoplankton, and more. As stated above, Bondavalli and Ulanowicz (1999) reported that American alligators in Big Cypress National Preserve in Florida (USA) confer net benefits on 11 of their prey including invertebrates, frogs, mice and rats. Thus, not only are indirect effects greater than direct effects in terms of the quantity of effect (linked to the real amount of physical energy or matter transferred), but also the dominance of indirect effects can lead to a reversal such that an effect that is negative in the direct sense becomes positive when all direct and indirect effects are integrated. Thus, for example, where an alligator predator gains at the expense of the frog prey in their direct interaction, the Frog (capitalized to signify a typical frog, or the collective role played by all individual frogs in the ecosystem) actually benefits in real physical ways (real flows of energy and material) from the presence and role of the Alligator in the ecosystem (capitalized again to the alligator type or collective alligator role in the ecosystem). In this case, the logic can be explained that the alligator also consumes other predators of the frogs (e.g., rats, snakes, etc.) such that absence of the alligator would result in more rats and snakes and fewer frogs, overall. In other words, this network approach allows one to simultaneously assess both bottom-up and top-down controls, an important area of ecological research.

We next examine a whole-system tendency related to ascendency that also serves to reconcile two seemingly opposing tendencies as is often seen in the dialectical nature of Life systems.

Lesson 6. Ecosystems Balance Efficiency and Adaptability. The robustness index of Ulanowicz shows that all ecosystem networks develop to maximize robustness of system operation by balancing (1) efficiency of flow (by pruning connections to achieve highly ordered structure) and (2) options for adaptation to disturbance (by maintaining alternative redundant pathways).

Ulanowicz (1980) used ENA methods based on information theory to develop the whole-system principle "ascendency" mentioned in Life Lesson 2. More recently, he modified the use and interpretation of ascendency to speak of a "propensity for ecosystems to increase in ascendency" (Ulanowicz, 2004) and to portray ascendency as one half of a balancing act, rather than ascendency itself being a goal function that ecosystems seek to maximize (Ulanowicz et al., 2009). In his original paper on ascendency, he wrote (Ulanowicz, 1980):

> The ascendency measures in one index the two attributes of system size and flow organization, i.e. it is a natural descriptor of the combined processes of growth and development. As we shall see, growth and development of flow systems are sometimes in conflict, and ascendency then serves to define the compromise configuration. (p. 227)

In the 2009 paper in which they presented the robustness index as a system principle able to depict this balancing act, he and coauthors wrote (Ulanowicz et al., 2009):

> Information theory (IT), predicated as it is on the indeterminacies of existence, constitutes a natural tool for quantifying the beneficial reserves that lacunae can afford a system in its response to disturbance. In the format of IT, unutilized reserve capacity is complementary to the effective performance of the system, and too little of either attribute can render a system unsustainable. The fundamental calculus of IT provides a uniform way to quantify both essential attributes — effective performance and reserve capacity — and results in a single metric that gauges system sustainability (robustness) in terms of the trade-off allotment of each. (p. 27)

And later, to explain an equation they presented showing that system capacity (C) equals the sum of system ascendency (A) plus system reserve capacity (Φ), they wrote (Ulanowicz et al., 2009):

... the capacity for a system to undergo evolutionary change or self-organization consists of two aspects: It must be capable of exercising sufficient directed power (ascendency) to maintain its integrity over time. Simultaneously, it must possess a reserve of flexible actions that can be used to meet the exigencies of novel disturbances. ... these two aspects are literally complementary. (p. 30)

Ulanowicz then calculated the ratio of network ascendency to network capacity:

$$a = \frac{A}{C}$$

and introduced an equation for robustness, R, similar to the Boltzmann equation for disorder:

$$R = -a \log(a)$$

Ulanowicz et al. use this equation to plot robustness ($-a \log(a)$) on the Y axis as a function of the degree of order (a) on the X axis. This robustness measure can be calculated using energy and material flow data for any ecosystem network and is closely related to system sustainability. All comprehensive ecosystem networks (with clearest results for networks with 12 nodes or more) studied thus far yield robustness values near the top of the curve showing the universal tendency for ecosystem networks to self-organize in ways to balance efficiency and adaptability (see Fig. 6.4). In fact, the distribution of the data for different ecosystems was so tight that Ulanowicz began referring to the optimum range as the "Window of vitality," a tendency for ecosystems to evolve that balances the efficiency—redundancy trade-off. This led to a spin off question of whether socio-economic systems would show a similar balancing act, and if not, how could they

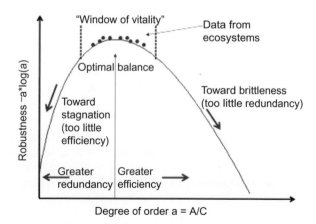

FIGURE 6.4

Redrawn from Ulanowicz (2009a) an early paper on the "balancing act" metric.

be designed and managed to display these characteristics. Recently, Ulanowicz and others have begun to study system robustness for economic and human food system networks (Goerner et al., 2009; Kharrazi et al., 2013; Fath, 2015; Fiscus et al., 2015; Fang and Chen, 2015). See Fig. 6.5.

In Fig. 6.5, we see that human economic and trade networks studied to date plot to the left of the robustness curve peak. This is most likely due to the fact that trade networks are more highly connected; and thus, show more redundant flow pathways than natural ecosystem food webs. The one data point for an industrial human food web—a subset of nitrogen flow in the US beef supply chain—plots to the right of the robustness curve peak. This is most likely due to the fact that this beef supply network is more highly pruned toward highly efficient linear flow; and thus, shows fewer redundant flow pathways than natural ecosystem food webs.

The working hypothesis of Ulanowicz, his colleagues, and ourselves is that the robustness index is generic and provides valuable insight about any flow network. Ulanowicz (2009a) provided sensitivity analysis methods by which to determine the amount that any individual flow link in a network contributes to

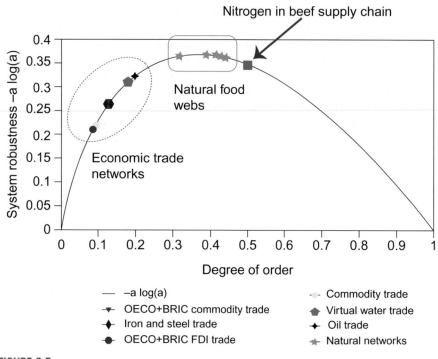

FIGURE 6.5

Robustness measure for various socio-ecological networks. Note, natural food webs cluster at the optimum trade-off between efficiency and redundancy; economic networks are overly redundant; and, the nitrogen in industrial beef supply chain is overly efficient.

overall system robustness. This measure enables one to identify those flows that if altered would most improve overall system balance of efficiency and adaptability. We also hypothesize that this universally observed network configuration pattern in natural food webs—systems which have evolved, survived, and self-sustained despite major disturbances over millennia—provides an excellent indicator by which to gauge the sustainability and robustness of human networks of many kinds (Goerner et al., 2018 submitted). This approach is the primary method driving a new effort called regenerative economics, which is led by the Research Alliance for Regenerative Economics (see capitalinstitute.org/research-alliance-for-regenerative-economics).

As for all the other Life Lessons and network science tools in this chapter, to obtain such systemic insight requires comprehensive data and holistic approaches such as ecological network analysis and NEA. We see these methods and the results they yield to support our call for more holism and synthesis in the foundational paradigm and daily working practices of science to understand Life. We have just one more Life Lesson, and it is one that may serve to fold many of the others together.

Lesson 7. A Hyperset Formalism of Life Prohibits Fragmentation of Life from Environment. We propose a novel science formalism to prohibit fragmentation of the Life—environment unity. This formalism employs hypersets to prohibit the fragmentation of Life from environment and to prohibit the fragmentation of three necessary and unified Life holons of organism, ecosystem, and biosphere.

As with much of the thrust of this book, the articulation of a need for a holistic formalism is anticipatory. We present a specific kind of science formalism to act now toward a specific future ultimate purpose to help solve the sustainability crisis in service to Life. In prior work (Fiscus et al., 2012), we explained the need for, and presented an initial version of, a science formalism to facilitate the win—win Life—environment relation while also promoting holism as balanced with reductionism. In addition to this anticipatory and problem-solving rationale for devising this formalism, we also see it to provide a better fit with reality—as for good, rigorous mainstream science, we assert this model has explanatory and predictive power, and we expect it to survive experimental tests.

Throughout this book, we have talked of the three unit-models of Life as each being a holistic, complex entity in its own right (unfractionable holon). However, we next propose a single "multi-model" (or perhaps a meta-model, a hypermodel, or a model made up of models) of Life as a fully unified whole integrating the organism, ecosystem, and biosphere holons.

Our approach uses hypersets as inspired by Kercel (2007) who built on work of Rosen on the need for impredicativity to model Life as we have mentioned. Kercel (2007) and Rosen (1991, 2000) invoked such methods to explain Life at the organism scale, but we also apply them to ecosystem and biosphere unit-models of Life, and to the task of unifying all three unit-models.

The use of hypersets requires self-reference and ambiguity (Kercel, 2007), two concepts which are difficult for other mathematical tools to handle. A hyperset is a set that contains itself as a member. Hypersets represent a more general class than regular sets, which are not allowed to contain themselves as a member; and thus, regular sets have no self-reference or ambiguity. Hypersets and what they contain can be nested to varying depths and display context-dependent ambiguity in terms of what they represent (Kercel, 2007). Kercel (2003, 2007) used hypersets to explain one logical resolution of the Liar's Paradox, to explain unique aspects of M.C. Escher's graphic art, and as a means to understand Rosen's metabolism-repair model for life. In Kercel's work, ambiguity is of central importance, and we employ this feature below.

We next present the initial development of this formalism only slightly modified from Fiscus et al. (2012). After that, we present more recent conceptual work to push this approach forward.

In the original work (Fiscus et al., 2012), we often spoke of the "community—ecosystem" as a holistic Life unit and of the "cell-organism—individual" as a unit. Here, we shorten these to ecosystem and organism, respectively, while keeping in mind the overlap with other terms and ideas for units of Life. Our starting place was to propose that something new is needed beyond the typical linear, nested ecological hierarchy that may be intuitive and conventional wisdom:

The biosphere contains ecosystems which contain organisms

Or, in terms of sets and borrowing Salthe's (1985) notation for hierarchical systems:

{biosphere{ecosystems{organisms}}}

This basic scheme has been used in many multi-scale conceptual frameworks for understanding Life's hierarchical organization. Barrett et al. (1997) and Rowe (1961), for example, presented excellent versions of this approach, again what is essentially a linear hierarchy. All of these have been useful for teaching ecology and environmental science, and for organizing research and research projects. But, as we know from the state of the world, they have not yet been successful as a conceptual model able to facilitate a healthy and positive human—environment relation.

Informed by the work of Rosen on complexity and the unique "closure to efficient cause" feature of organismal life he identified, and the work of Kercel on impredicativity and hypersets, we sought a new representation and model in which *the relational hierarchy forms a closed loop.* This can occur if the top level of the hierarchy links back around to the bottom level, so the whole forms a continuous hierarchical loop. Kercel (2007) wrote of a "closed loop hierarchy of containment of entailment" to describe Rosen's metabolism-repair model.

To move the familiar linear hierarchy to a closed loop hierarchy, we can start by describing Life (Life itself, Life as a unified whole, the Life—environment system) in words using the idea of containment, and adding the closure twist:

Biosphere contains ecosystems contain organisms contain biosphere.

Or in set notation:

{biosphere{ecosystems{organisms{biosphere}}}}

This is interesting perhaps, but not quite right, so we modify it again. Following Kercel's lead, we employ ambiguity—or we could say a context-dependent identity—for what we mean by "biosphere" in the above closed loop hierarchy.
In 2012, we wrote:

> If we take biosphere to mean the entire planetary environment on one hand (at the outermost level of the hierarchy) but also to mean any physical, chemical or molecular subset of that planetary environment on the other hand (at the innermost level of the hierarchy), then the closed loop hierarchy can make sense and be of great use. Organisms and cells do not contain the biosphere in terms of encompassing the entire planet inside their spatial boundaries. Instead, organisms and cells contain abiotic, physical—chemical, environmental elements and molecules necessary for life, each of which is of the same abiotic, physical—chemical nature as primary elements of the planetary environment.

Substituting "environment" for "biosphere" to make it clearer, we could depict Life using the closed loop hierarchy this way, with another hyperset:

Life = {environment{ecosystems{organisms{environment}}}}

The potential benefits and ultimate implications of this holistic, self-referential multi-model of Life are profound. These benefits and implications are the same ones we have touted as the focus and need for this book—to change our system of ideas and then heal our relationship with our environment. The above hyperset equation needs more explanation, development, contextualization, and corroboration, but we see it as opening a new path and vista for holistic modeling of Life—environment.

Consider this quote which may help explain the two inner most levels in the hyperset equation and hierarchy, "organism" and "environment" inside it. Olomucki (1993, cited from Lahav, 1999: 62) wrote of the necessary ambiguity and complexity well:

> When we attempt to define life, or living, we immediately come up against a fundamental and apparently irreducible paradox: living organisms are composed of inanimate molecules. . . . Must we then say that 'life' is the interaction of all the inanimate components of this whole? In other words, that nothing is alive in a cell except the whole of it?

In the hyperset model, as well as conventional wisdom, "environment" has an inherently and perhaps predominantly "abiotic," nonliving, inanimate, physical—chemical nature. Water, water vapor, clouds, and rain, for example, as configurations of H_2O molecules in and of themselves, are not alive. But, at both the outermost and innermost levels of the hierarchy, these abiotic elements, components, laws, forces, and features of the physical–chemical environment have come under the influence of Life in ways that benefit Life in both its discrete and sustained forms. We can think of multiple examples of the "inner environment" inside organisms and cells that are primarily of physical–chemical nature—pick any collection of basic molecules. And, we know from the thought experiment of tracing individual carbon or oxygen atoms (as in Chapter 5) that the major Life elements flow and cycle continually between organism, ecosystem, and biosphere realms. Furthermore, in a relation similar to autocatalysis, each of these Life holons "feeds," augments, enhances, and improves the others. The higher redox potential of the oxygen atmosphere of the "outer environment" became integrated into the inner most metabolic processes of all aerobic cells and organisms. The ecosystem holon in the middle, with its food webs, nutrient cycling, and radical network interdependence, serves to channel the materials and energy of dead organisms into improved soils and future generations of discrete Life forms. As a unified system, Life–environment improves both Life and environment over time.

If this proposed formalism stands the tests of critique and scrutiny, then it would depict a relational closure that fully unifies all three Life levels. Viewed and modeled this way, organisms, ecosystems, and environment (or biosphere) are not separable or independent. The inherent paradigm shift would mean that all prior fragmentation of Life—perhaps a series of useful analytical simplifying assumptions that became *reified* or taken to be actually true rather than consciously chosen models (since no synthetic counterpart process qualified it)—would be repaired. Such fragmentation would also be explicitly prohibited going forward. The primary mode of action by which this formalism helps is that by prohibiting any fragmentation of Life from environment, this *prohibits any devaluation of environment* relative to Life. If we understand the value of environment—as essential Life support and context—to be primary, then the scientific paradigm, research, technology, applications, and policies can follow to protect and realize this value. If we do, then we fully expect better outcomes, as there is literally no possibility for any "Tragedy of the Commons" to occur.

As supported by our work, the works of many we have cited, and we hope concisely with the hyperset formalism, Life cannot be meaningfully identified or defined at any single scale. With these others, we see the Life–environment system as inherently multi-scale, requiring a multi-model, complex in the sense of Rosen (containing impredicativities), and unfractionable in the sense of Rosen (components cannot be understood, analyzed, or managed independently). This unfractionable nature of Life is synonymous with the wholeness of "organic systems" of Life as depicted by Ulanowicz (2001).

In Chapter 5, we saw that Rosen asserted that organisms are self-defined, self-causing, and self-making during a single discrete lifespan. But, for Life to persist long term requires an inherent capacity to bring new self-making organisms into existence in perpetuity. This additional, "trans-organismal," and self-referential functional relationship we referred to before as the "self-making of the self-makers." This fundamental concept of a closed loop hierarchy can also be seen in food webs and trophic levels—"top level" predators die and their material bodies are recycled back through the "bottom level" soils and decomposers and pass through and feed plants and thus around the biogeochemical cycle, ad infinitum. It is important to note that this sustained Life capacity requires the abiotic, physical–chemical environment as an active participant, and, for truly open-ended and perpetual existence the environment must *improve* over time.

This model, "simple" and brief to write down, holds potential to change much. The dictionary definition of Life, views of the origin of Life, our shared cultural mental model of Life, the paradigm of Life science, the values of Life and especially environment, policies, actions and behaviors as they relate to Life, sustainability of human life—all these can and should change for the better. We believe this approach works objectively in that it is compatible with and better explains empirical evidence and real observations. And, we believe it works subjectively and in anticipatory fashion since it aids human change toward ideas and actions to enhance Life and its necessarily integral environmental context. This model is also falsifiable—any observation of a single organism, species or functional type self-perpetuating in isolated relationship with the environment would potentially falsify this hyperset model. Thus, we plan to examine closely the work of Chivian et al. (2008) and the "single-species ecosystem" they report to have found. Such a finding would collapse the second and third holons to one, but still would not overcome the self-referential aspect of the environment. A new hyperset model, assuming there is a single cell ecosystem, would be

Single cell ecosystem Life = {environment{ecosystem-organism{environment}}}

To develop our hyperset model beyond the 2012 introduction, we have explored existing mathematical approaches that are potentially compatible and mutually supporting. For this effort, we have sought mathematical tools with any of these features: recursion, self-referential, holistic, cybernetic, complex, generative, relational, and multi-scale. We identified a set of five associated mathematical formalisms and have begun to develop methods to integrate these. The allied holistic formalisms, many of which we have already discussed, are as follows:

1. Fractals, which arise from recursion yet remain bounded
2. Network models used in ENA, including
   a. Autocatalytic loops of Ulanowicz
   b. Network enfolding of Patten
3. Category theory and Rosen's unfractionable metabolism-repair model for life
4. Impredicative logic—as associated with hypersets and as represented in graphs by Kercel
5. Graph theory—to represent hypersets in a graph or network construct

We plan to develop necessary paths and intermediate steps to integrate hypersets, fractals, networks, category theory, and graph theory as applied to the hyperset formalism for Life–environment above.

In future work, we plan to make the model more rigorous and testable, to generate testable hypotheses, and to propose experimental tests. Next steps include simulations, visualizations, and identifying natural systems that can serve to test the hyperset formalism. We are hopeful that this holistic model, along with other progress in holistic Life science, will allow us to manage the unified Life—environment as a systemic whole to achieve true and systemic environmental sustainability.

SUMMARY AND SYNTHESIS OF CHAPTER 6

Referring back to where we started, in Fig. 1.3, we presented a simple cartoon model with two instances of human self, world environment, and the relation between. One of these instances is the "real" (physical, ontic) world—how we humans actually exist, what our planetary environment is really like, and how the relationship occurs, manifests, and unfolds. The other instance is a mental (epistemic realm) version of the same triad—as in the thought bubble, we also have our ideas, beliefs, worldviews, mental models, and science of self, world, and self-world relation. Despite the cartoon depiction, we assert this internal arena is the primary place we need to examine, work and change to solve the crisis we now face. And in this chapter, we have shown seven lessons, tools, methods, and conceptual approaches that provide pragmatic extensions of a new way of thinking and help corroborate the need for fundamental reform in science.

Another ally for this work is Frances Moore Lappé, including her book *EcoMind: Changing the Way We Think to Create the World We Want* (2011). She similarly identified cultural mental models and mindsets as pivotal for environmental sustainability. She described seven "thought traps" as part of the "dominant frame" of conventional wisdom—including views within the "environmental movement"—that cause and continue to enable environmental degradation. Much like Donella Meadows, who identified the paradigm and the power to transcend paradigms as our most powerful forms of leverage for change, Lappé wrote how a mental shift can yield access to new forms of power:

> In the dominant coherent, yet self-defeating way of seeing, the "environment" is something outside of ourselves that needs help, really fast. From this standpoint, one perceives oneself as joining and enlisting others in an environmental movement to rescue the planet.

> But as we rethink the premises underlying this worldview, we move to a different place altogether — a place where we experience ourselves and our species embedded in nature. (p. 16)

And a bit later:

> With an eco-mind, we move from "fixing something" outside ourselves to re-aligning our relationships within our ecological home. (p. 16)

She provided an excellent book with many insights for mental shifts that can yield tangible improvement. She explained how our species, with its capacities for imagination and creativity, are very much capable to create the world we truly want—if we are able to examine and challenge some of our most closely held ideas. We mention this in conjunction with recognizing that humans exist primarily at the organism holon of our hyperset model, but currently are the dominant driving force in all three holons in most locations on Earth. That model again is

Life = {environment{ecosystems{organisms{environment}}}}

We are the dominant force altering the planetary environment and biosphere—this has led to the term "Anthropocene" to describe the current unique epoch of Earth history. And, in most ecosystems, such as local communities, watersheds, and regions, we also shape most of the energy and material fluxes and processes. Thus, human imagination and creativity are primary forces in every aspect of this holistic Life—environment system.

Lappé's general solution, "to think like an ecosystem," fits with the conscious choice to form, envision, and operate with holistic mental models of self-sustaining Life forms and Life—environment relations, as we have proposed. Lappé (2011) also wrote about how this is very different from a mechanistic and reductionist worldview. In Chapter 7, we return to work of Robert Rosen for strategic guidance on how to bridge from science to applications and technology, keeping his critique of the machine metaphor in mind.

A bridge not too far: Spanning theory to science to application

7

INTRODUCTION

In Chapter 6, we described seven lessons learned about Life and the Life—environment relation gleaned from decades of holistic science using network analysis and related methods. We also advocated the view of science as an active participatory process with multiple impacts on the systems of study and the world environment as a whole. Conversely, we urged against any misconception of science as a passive one-way process of information moving from Nature into our minds in the form of new or refined knowledge. As we move to discuss and recommend technology and applications for sustainability, we will again promote a view toward a complex, multiway flow, and holistic interdependence between four realms—science, technology, culture, and the world environment. In this chapter, we present ideas we see as crucial for understanding the bridge or conduit—perhaps even better seen as a looping, cyclic, or spiral pathway—between the holistic Life science we have described in past chapters and the applications, technologies, and solutions we examine in Chapter 8. We referred to the "root metaphor" and "system of ideas" of industrial culture in Chapter 1, and in this chapter, we examine the identity, roles, and implications of the root metaphor and system of ideas we create, promote, and employ.

The two most important concepts we develop are:

1. Both the "encoding" (like reading in, observation, and measurement) and the "decoding" (like writing out, predictions, and experimental tests) modes of Rosen's modeling relation and model for science are always operating in concert. Rosen also described "fabrication" as this relation running in reverse—to start with a formal model in mind and then realize that model in material form in the real world. Rather than separate sequential processes of forming and refining science knowledge vs the fabrication or realization counterpart—we assert that this is a single integrated process with interdependent two-way flow between the science/mind/idea and physical/material/environmental realms. Science and culture, like Life, are dual operating as receiver and transponder (as Patten's Janus hypothesis, Patten, 2016b) and we make a mental model of this reality.

Foundations for Sustainability. DOI: https://doi.org/10.1016/B978-0-12-811460-5.00007-8

2. Idea 1 can be generalized as *our mental models change the world*. Next, given the dominance of the machine metaphor as root metaphor in both our science paradigm and basis of industrial culture, we propose that we are in essence *making the world into machine*. We see industrial science, technology, and culture as actively altering the world to fit our preconceived image of a fundamental mechanistic unit of reality. This in turn helps explain and provides a coherent story for how and why we see the most essential Life support services of the world, the fundamental integrity and health of Life itself, breaking down, wearing out, "running out of steam" (or gas)—symptoms which only apply to machines and make no sense as behaviors of living systems which normally do the opposite. We have seen in this book how all Life systems self-organize, self-repair, self-improve, and self-sustain, as long as we consider them holistically as integrated organism, ecosystem, and biosphere holons within the Life—environment holon.

We next recap and expand our discussion of Rosen's modeling relation. We also review Rosen's characterization of the machine metaphor (which he traced to Descartes) and its inherent problems. We highlight corroborating work from Hornborg, Ulanowicz, Goerner, Lappé, and others who support and provide further insights into the power and implications of this metaphor as well as aiding the creation of a better metaphor. We frame options for what an alternative root metaphor—or a suite, set, or system of root metaphors—would be like. Finally, we segue to how these ideas can guide and inform the qualities we need most in applications, technologies, policies, and related actions.

The lens through which we now view the need for an integrated bridge (such as a spiral staircase) and interlocking interface between science and applications (technology and culture) fits with our start in Chapter 1. There, we mentioned the main book idea to help reform the closely integrated system of ideas in culture and the paradigm in science. Here, we learn more of how and why mental models are so powerful, and by extension how powerfully they serve as sources of leverage for change. A key set of working assumptions for our next steps are stated below.

We assume (hypothesize) that the current root metaphor, main theme, narrative, motivating idea of human self, world, and self-world relation have been infused *by science* with ideas about machines and mechanisms. We adopt the view, as Rosen, Ulanowicz, Hornborg, and others have shown (and as they show more below), that use of the metaphor of mechanism as a fundamental unit of how the world works—applicable to Life, humans, and natural systems—starting from Descartes, Newton, and others hundreds of years ago—determines many critical aspects of how we do science, how we do technology, and by extension how we think and live, and how we get the world we ultimately end up with.

Of the many details we can study on the impacts and implications of mechanistic science, the main issue we treat in this book as "actionable" for transformative change now, is that by treating Life forms, humans, and the world as mechanisms, we have come to expect that we and our world will necessarily wear out and break down. This is in essence a kind of modern industrial "world myth" that all machines, all Life, and all of existence in the universe are headed down the ultimate one way, dead-end pathway required by the second law of thermodynamics (which mandates the tendency toward increased entropy for isolated systems).

One additional ramification of this dominant mechanistic and *entropic* narrative we read in mainstream science and integrated industrial culture is that, since environmental, world, and even universal decay and degradation are treated as inevitable and absolute (scientific law), it then is acceptable and to be expected— in fact, one has no choice at all—that the impacts of day-to-day science activities, campuses, and enterprises must also lead to decay, degradation, and dissipation of higher quality energy and resources as the world, and all actors in it, march to inevitable heat death. We note the daily activities and operations of science here first, as we want to focus on the integrated paradigm-action of science as both source of our problems and our solutions. But, the same assessment of degrading impacts applies to all other industry, commerce, development, transportation, etc., in our industrial culture. Each and every sector is "allowed" or assumed to "do work," contribute or aid progress at the expense of exporting entropy and leaving the environment depleted, exploited for energy and materials, and thus worse off by necessity.

A final word on this hypothetical and roughly sketched science-culture plot and story line is that the overemphasis within mainstream life science on the "life as organism" model plays a critical supporting role. Just as all machines wear down, and any isolated physical system winds down, all organisms die. After early growth and development, mature organisms are, like machines, in a constant process of senescence, aging, and loss of physical vitality. Thus, the shared cultural story of the fate of self and world, and of Life and environment, are thematically consistent and unified—both must decay and die. And, the relationship between them is thus fittingly colored by mutual influences of harm, threat, damage, and antagonism. Given this individualistic cultural story, and the ultimate survival stakes, it is "us against them" as we humans and other Life forms relate to the environment, to each other, to other species, and to the world as a whole.

As we have seen and will continue to develop next, this linked system of science paradigm, root metaphor, "world myth," and the culture and technology they drive has two strange aspects currently arising:

1. The real evidence of environmental decay and threats to essential life support services provides a stark wake up call for us to examine our subconscious, unconscious, or collective (perhaps buried, repressed, or walled off by fear or economic pressure) belief systems (ideas, epistemic realm). One way to interpret this is to ask if we have gotten ourselves trapped in a kind of self-fulfilling prophecy. And then, we can ask if this is really what we want, and if this is really the only way? Can we not do better? Or, even define differently,

what "better" means in a Life–environment worldview. This trap—perhaps a form of "cognitive dissonance"—relates to the internal mental models (our mindsets) as in Fig. 1.3.

2. Real evidence of nonhuman systems (and some nonindustrial human systems) shows clearly that the environment and Life in its ecosystemic and biospheric modes need not degrade over time—the oxygen atmosphere, continually improving soils, high biodiversity, creation of those carbon-rich deposits that yield fossil fuels, and other real effects show that a nonentropic progression is fully possible and perhaps the norm for Life (Life as a unified whole). This relates to the real-world instances of self, world, and relation between in Fig. 1.3.

This part of the book sheds additional light on our critique so far of what an unbalanced, inaccurate, or inappropriate science paradigm, root metaphor, and linked techno-culture can do. We can see ideas spread via the modeling relation and science-culture conduit to infuse the whole world with our most closely held, most often revisited, and continually manifested images of ourselves, Life, the environment, and how the world works. If this organic unity of mind and nature, an idea we adopt from Bateson (1988), has worked to cause harm (an unintended consequence of the machine metaphor which has produced many great gains), then with a new image in mind, it can now begin work to heal, repair, and regenerate Life systems.

ROSEN'S MODELING RELATION REVISITED—A DOUBLE-EDGED SWORD

Rosen criticized the machine metaphor, the dominant working model for science, because he saw it as limiting and even dysfunctional for his most passionate quest to understand What is Life? or answer the question, Why Life? His synthetic idea to identify the machine metaphor as culprit blocking his progress is closely tied to our main thesis of this book:

> The dominant mental model of the mainstream science paradigm works to alter the environment in detrimental ways and blocks progress toward true sustainability.

Borrowing again from Rosen, we go more in-depth here into how this relationship fits with the modeling relation he used to represent and understand the process of science.

As noted in Chapter 6, Rosen wrote that the encoding and decoding processes of his modeling relation, as well as our sensory and perceptual apparatus we use to encode and decode, sit between natural and formal systems, between the self and the ambience of the environment. He wrote that encoding and decoding are ambiguously part of both natural and formal system realms. This ambiguity or dual-nature allows these two portions of his modeling relation to serve as a conduit or interface by which the formal systems of individual mind (and collective science) act to mold the material world. Via this mode of operation of the modeling relation, our ideas become manifest in real, physical systems.

Rosen (1991) described the interface not only as active but also with a kind of inherent flaw, repeating one of his quotes:

> ... what we observe are *phenomena*, which are to an equally unknowable extent corrupted by our perceptual [sensory and mental] apparatus (which of course sits partly in the ambience). (p. 56)

The "corruption" aspect is linked to the common philosophical dilemma that we cannot perceive or know anything about the natural world in any perfect, noise free, or exact way. This is also related to the mind–brain and mental–physical problems—we still have no single, unified, consensus theory of the interface between the organic, material, biological construction of our brains and bodies and the abstract seemingly immaterial ideas, symbols, words, and images they can conjure, create, manipulate, and communicate. The quote gets at the recursive aspects we try to capture in the hyperset equation of Chapter 6 because we reside in the same environmental ambience that we try to understand and formulize. Practically, this dilemma appears in mathematical modeling of ecological systems, in that the perceptual apparatus is different for all investigators, and thus, the conceptualization of the phenomena into quantifiable systems will differ markedly based on the observer. This is even referred to as the modeling problem, and there is surprisingly little theory to deal with this foundational problem. Yet, models are made, often fragmented in structure and function; they remain useful to solve immediate problems, while they typically fail to address the deeper root problems. As we show below, Rosen has probably gone farther than most in addressing this. While it is impossible to know the unknowable or avoid completely the corruption, addressing the core issue of fragmentation in the current systemic human–environmental crisis may yield new insights to help understand the perennial mind–brain question.

We already presented Rosen's modeling relation in Chapter 6, but we need to examine more the various modes of action between formal and natural systems. Figs. 7.1 and 7.2 will help.

In Fig. 7.1, we changed the depiction of the formal system to be like a tank or reservoir that can increase in content as knowledge is gained. We also added a main flow of action arrow moving from left to right, and we changed the caption to describe the mode of operation of doing science, or systematic human learning about the natural system, which represents the real world. While the caption suggests that science starts with step 2, encoding as in observation, we could also start the process with step 4, decoding as in making a prediction, or with step 3, inferential entailment as in forming a hypothesis. Wherever we start, progress proceeds counterclockwise in this simplified, yet unfragmented, version of how science gains knowledge of the natural world.

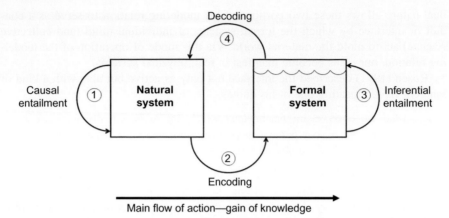

FIGURE 7.1

Rosen's modeling relation modified again from our Fig. 6.1. In Rosen's representation of the scientific process, science operates with a main flow from left to right in clockwise steps and can be thought of as starting at step 2, encoding, which is the realm of measurement and observation. As work proceeds counterclockwise, the formal systems (science models, theories, etc.) are successively and progressively formed, refined, and improved and knowledge increases. Thus, the formal system is shown as a reservoir that can gain in content.

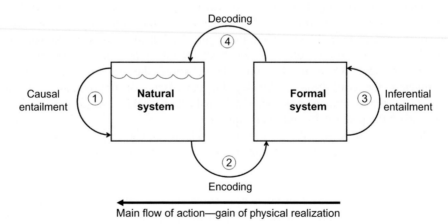

FIGURE 7.2

The process of "fabrication" modified after a diagram of the modeling relation (Rosen, 1991). The main flow of action now is the gain of a physical, material realization—a new construction in the natural system, the real world, based on an idea, mental model, or design in the formal system.

A related idea we see as crucial and profound is that in the standard science process, the formal system is assumed to change substantially (much new knowledge is gained, the collective reservoir of science grows deeper) while the natural system is assumed to change minimally, if at all. This assumption is linked to the foundational mainstream science tenet of *objectivity*, which we examined in Chapter 5 (where we discussed the alternative principle of internalism).

We think this is a fair assessment of the mainstream model and daily operations of science—progress made by increased knowledge is essentially the sole focus of the totality of the social and cultural science enterprise, and the individual daily actions of the vast majority of all scientists. The impact of the science process on the natural system, on the real world, is ignored and treated as either nonexistent or as an "externality" in the economic sense, arbitrary external impacts that do not enter into the focal accounting or calculus of costs, benefits, expenses and gains, and profit and loss in both monetary and knowledge realms. Scientists work to construct experiments and studies to be objective—as if their "corrupt" sensory, perceptual, and motor apparatus (of encoding and decoding) could interact with the natural system while leaving it completely intact and untouched.

A related idea is that the natural system is infinite and inexhaustible, both in terms of physical material aspects (such as energy, resources, and space to absorb human industrial wastes. See Chapter 5 about the empty world metaphor of Goodland and Daly, 1996), but also in terms of the knowledge of the world, all that could be known, myriad, and unexplored aspects of reality. Thus, in Fig. 7.1, the natural system has no "fill level" indicated—gains or losses, increments, and accounting are not considered.

What we are saying is also needed, and what Rosen suggested, is seeing that the flow, relation, mode of action, and impact *is always both ways*. And, we assert that conscious awareness of this dual action is critical to success with human−environment sustainability.

Rosen wrote about two separate paths in his diagram, and the numeric step labels still apply to Fig. 7.1:

Path 1—causality in the natural system. This is how the world works.
Path 2 + 3 + 4—the integrated process of scientific knowledge generation and testing.

Rosen (1991) wrote that the ideal is for this modeling relation to "commute" (obey a commutative property) such that

Path 1 = Path 2 + 3 + 4

If the inferential entailment, with its associated formal system (model), does commute—that is, if we can successfully decode it back to the real world (make predictions) and successfully encode it again (observe that the predictions held true)—then by one or more cycles of this process, we have built and confirmed a reliable representation of the natural system.

Rosen noted that one key power gained from successful use of this modeling relation and science process is that we are able to learn about the natural system by asking questions about the formal system. Another key power we see to be gained is that we are emboldened to believe that we understand how the world works and we can create new models, new designs, that once constructed and realized in the physical material world will work properly, will function, and operate successfully to achieve the goals and ends we have designed into them.

We want to examine more of what Rosen wrote about this process operating in both directions. First, we look again at what he said of the ambiguous nature of the links between formal and natural systems. He wrote of the encoding and decoding arrows (Rosen, 1991):

> The encoding and decoding arrows in this case are still *unentailed*, but it is no longer clear *how* they could be entailed, or from what. These arrows are not part of the natural system, N, nor even of its environment; although they pertain to the ambience, they do not belong to it. Neither do they belong to the formal world of the self either; they look like mappings, but they do not compare formal objects; hence they cannot be mappings in any formal sense. Thus these arrows, which play the central role in comparing causal and inferential entailment ... turn out to possess a new and ambiguous status, equally within, and outside of, both the self and its ambience. (p. 62)

He wrote about encoding as "associated with the notion of measurement" and as "encoding phenomena in N into propositions of F." And decoding as "predictions about N" and also "decoding propositions of F back to phenomena in N" (Rosen, 1991, pp. 59–62).

His comment in the quote above—that encoding and decoding "possess a new and ambiguous status"—becomes important below.

Rosen (1991, 2000) wrote of "fabrication" and how this is the realm of life. This is the complementary half of the modeling relation as a two-way process. One can make a model of a natural system in a formal system, and/or one can realize a physical system based on a formal model. The latter route is fabrication, and he went so far as to say that when we practice biology as based on his relational theory (Rosen, 1991):

> ...we enormously enlarge the scope of biology itself. Biology becomes identified with the *class of material realizations* of a certain kind of relational organization, and hence, to that extent divorced from the structural details of any particular kind of realization. It is thus not simply the study of whatever organisms happen to appear in the external world of the biologist; it could be, and in fact is, much more than that. Biology becomes in fact a *creative* endeavor; to fabricate any realization of that essential relational organization (i.e., to fabricate a material system that possesses such a model) is to create a new organism. (p. 245)

Rosen's big idea here is that the modeling relation, which he says is the essence of science, also works as a conduit or transformation route by which ideas (in the form of relational models) become realized in the physical world. He speaks of fabrication and creative endeavors when the model is his unique relational one (with the "closed to efficient cause" feature) and suggests that one can create life forms—"create a new organism" as he says. But what happens, or what possible things can happen, if the formal system in mind is a mechanical model rather than his complex relational model of life? *What happens when we fabricate or manifest in the world realizations of the machine model that we have in mind?*

OUR IDEA—FORMAL SYSTEMS (SCIENCE, KNOWLEDGE) AND NATURAL SYSTEMS (FABRICATION, TECHNOLOGY) ARE ALWAYS CO-CREATING EACH OTHER

One obvious answer is when we consciously and intentionally design technological machines, devices, or tools, we first conceive of the design in mind (a model, or formal system, aided by mathematics, life sciences, physics, and other science and engineering methods) and then build a prototype and eventually a finished machine or device as a real physical, material object. For this standard process of technological development, we may traverse the modeling relation many times—building the device, measuring its performance, learning more, changing the design, building the new version, testing performance yet again, and so on indefinitely. And, at times, the wheel of the modeling relation cycle spins the other direction—the technological development process can go onto a tangent of more research, new learning, and gaining new science knowledge on a path inspired by something learned on the journey to develop a new device.

However, we propose that this realization of a machine model can also occur *unconsciously and unintentionally*. We can unintentionally realize a physical manifestation of the machine metaphor which dominates the formal system (science paradigm) *when we forget that the machine metaphor is just an idea of our own construction*. What started with Descartes and Newton was an analogy or useful simplifying assumption—what if we treat the world as if it were a clock or mechanism? But, this seems to have become *reified* or ossified as in a hard habit in the sense that we skip the "what if . . ." and "as if it were" parts and short-circuit this so as to "treat the world as a clock or mechanism." This becomes less like encoding and decoding, prediction and testing, and more like *projecting, fabricating, and actuating*. This is more like considering the one-way flow of Fig. 7.2, but we urge considering both flows simultaneously.

In Fig. 7.2, the natural system is portrayed with reservoir status—it can gain in content of new inventions, devices, and technologies. Here, the formal system, exaggerated to the extreme of only representing a single dominant machine metaphor at the heart of the current science paradigm, is infinite and unchanging. The

machine metaphor is inexhaustible and seems universally applicable in every context and for every need. It can be used, employed, realized, built over and over and it never gets depleted or degraded—or even challenged—due to the absolute way in which it is trusted and habitual throughout all corridors of science, technology, and culture.

The unconscious or unintentional use of the machine metaphor we have described can be seen in many examples and cases such as the way we treat land during conventional industrial agriculture. A recent article in science explained that plants could be "tricked" to grow faster and longer under artificial light. The manner of presentation was only positive that this may be a good breakthrough to increase food supply for a growing human population, but without any reflection that it further treats the plant as an object or tool to be manipulated and industrialized, without any context of how that plant fits into the ambience of its environment. We explore the case of agriculture and others below in the section of this chapter on "We Are Making the World into a Machine." Before that, we see what others have to say about the machine metaphor.

THE MACHINE METAPHOR—MORE VIEWS ON THE LEGACIES OF DESCARTES AND NEWTON

While we have relied heavily on Rosen in this book, he is joined by a chorus of harmonious thinkers and researchers in a variety of fields in his focal critique of the machine and the machine metaphor as a crux issue for understanding and solving the sustainability crisis. Alf Hornborg, a cultural anthropologist also working in human ecology, wrote a book, *The Power of the Machine* (Hornborg, 2001), in which he examines the machine metaphor and related ways that machines and technology figure into the cultural as well as material dynamics of the environmental crisis, social inequality, economic inequity, and sustainability. One of his strengths is his interdisciplinary training and research in natural sciences, cultures, and social sciences. He presents an authoritative perspective that corroborates Rosen's and our statements above that science and the realm of ideas and models, and the natural, physical, material world, are inextricably entangled. He wrote about how this can also be confusing to sort out and described the focus of his book (Hornborg, 2001):

> In very general terms, I argue in this book, the problem is our way of conceptualizing the relationship between sociocultural constructions and material processes. We seem to have difficulties understanding exactly in which sense human ideas and social relations intervene in the material realities of the biosphere. Rather than continuing to approach "knowledge" from the Cartesian assumption of a separation of subject and object, we shall have to concede that our image-building actively participates in the constitution of the world. Our

perception of our physical environment is inseparable from our involvement in it. (p. 9—10)

Hornborg goes into great detail to reveal hidden, "mystified," and reified aspects of machines, technology, and unequal exchange of natural resources including a major critique of ideas about money and the current predominant money system in industrial culture. We discuss money in Chapter 8 and Chapter 9, but for now focus on the inseparable and interdependent relationships between the "image-building" of our science process and involvement in the physical biosphere. Hornborg (2001) also wrote:

> If, since Newton, the machine has served as a root metaphor for the universe, an advocate of a less mechanistic worldview might begin by demonstrating that even the machine is an organic phenomenon. (p. 10)

Note that he did not say the machine is "physical" or "material" but *organic*. We interpret this adjective as his recognition of the necessary role of humans in the idea, design, construction, operation, and use of any machine, as well as the interdependencies of machines and technology with living systems and natural resources of the world. He also wrote of his book and its title:

> If the word "power" in the title is ambiguous, so also perhaps, ultimately, is the word "machine." My concern with modernity and the social consequences of abstraction finally addresses modernist rationality as a *machination* in the widest sense. Although my primary argument is that the machine is social, it is embedded in reflections about the inverse observation that our modern social system functions like a machine. (p. 5)

Here, Hornborg echoes our views above that modern abstraction—such as modeling, scientific inference, and related processes—has social consequences and that we see human culture (with science and technology playing lead roles) operating like the machine he, and we, identified as the root metaphor. He also helps to show how the social system of industrial culture, as machine, can directly explain environmental and social degradation (Hornborg, 2001):

> The sum of industrial products *represents* greater entropy than the sum of fuels and raw materials for which they are exchanged. The net transfer of "negative entropy" to industrial centers is the basis for techno-economic "growth" or "development." In other words, we must begin to understand machines as thoroughly *social* phenomena. They are the result of asymmetric, global transfers of resources.

A cynic might conclude that the growth in the developed, industrial centers is predicated on the extraction, transfer, and continued impoverishment of the resource rich, yet industry poor, regions. In other words, the asymmetry is a design feature and not an unintended consequence. Can recognition of this connection and fairer valuation of those resources help mend this fragmentation, injustice, and unsustainable relationship? Soon after, Hornborg continues:

> Inversely, the non-industrial sectors experience a net increase in entropy as natural resources and traditional social structures are dismembered. The ecological and socioeconomic impoverishment of the periphery are two sides of the same coin, for both nature and human labor are underpaid sources of high-quality energy for the industrial "technomass." (p. 11)

As we proposed above in this chapter, Hornborg connects the supporting role of entropy and the second law of thermodynamics, other key pillars in the mainstream science paradigm, in this shared cultural story that accepts environmental degradation as unavoidable as well as acceptable in return for the benefits of industrial technology. This defeatism turns a technomass into the "technomess" that we have today. We return to Hornborg below when we discuss alternative root metaphors, and later when we address the role of money in these processes, but we next hear more views on the machine metaphor.

We have studied in detail how Ulanowicz showed autocatalysis and indirect mutualism to refute the validity of a monolithic mechanistic model for living systems and nature, to challenge the primacy of competition in neo-Darwinian evolutionary theory and to indicate the need for his ecologically-based root metaphor in the science paradigm. We have seen how Ulanowicz developed an ecological metaphysic as an alternative to the prevailing machine metaphor in science and culture. Prior to developing his three tenets of the "third window" he recommended as a needed improvement beyond the dominant vistas of Newton and Darwin (Ulanowicz, 2009b), he had previously challenged five founding pillars of the mechanistic paradigm of science. He showed these five working assumptions as unable to provide a coherent framework for understanding Life in its ecosystem and network forms. He wrote (Ulanowicz, 1999a):

> If one wishes to understand the development of biological systems in full hierarchical detail and is not content with the abrupt juxtaposition of pure stochasticity and determinism found in neo-Darwinism (Ulanowicz, 1997), then one must abandon the assumptions of closure, determinism, universality, reversibility and atomism and replace them by the ideas of openness, contingency, granularity, historicity and organicism, respectively. That is, one must formulate a new metaphysic for how to view living phenomena.

The five main concepts that Ulanowicz (1999a) proposed serve to differentiate the Cartesian, Newtonian, mechanistic metaphysic of mainstream science from his ecological metaphysic that is compatible with our holistic Life science in this book. This is how we summarized his alternative pillars of science (Fiscus, 2007):

> His *openness* refers to ontic or causal openness and suggests that chance is real and active, not merely a source of "noise" or "error," and thus not all is determined or determinable. His *contingency* relates to that qualified answer so often heard in ecology, "it depends...." Most if not all events arising from

cause-effect relations are not static but are contingent on other outcomes, relations and the context in which they occur. *Granularity* is his antidote to unrestrained universalizing. A granular extent to a science law, model or principle would admit limits to domains of applicability, would seek to "renormalize" or reconsider frames of reference in vastly different systems, and would be compatible with locally unique forms of place-based science institutions as suggested for locally environmentally sustainable science. *Historicity* is irreversibility and the importance of the time course of events – what happens and when it happens both matter, and many processes show hysteresis and do not run the same way backward as forward (e.g., soil wetting vs drying). *Organicism* is the operating assumption of unfractionability and wholeness of living systems, the opposite of or counterbalance to reductionism. (p. 52–53)

In this and other works, Ulanowicz helps to provide more nuanced understanding of what we have been referring to as the machine metaphor. He makes the distinct point that mechanisms *are* found in living systems, or we could say that a mechanistic model *is appropriate* in many cases for understanding nature. His objection and his reason for developing his ecological metaphysic are due to his sense that mechanisms are not able to tell the full story and are insufficient to explain Life as whole.

Ulanowicz accepts treating the first two Aristotlean causes—material and efficient cause—as mechanical forms of agency. He describes mechanical behavior as those cases that are deterministic, rigid, and inflexible: "If A, then B—no exceptions!" (Ulanowicz, 1997) Real processes in natural systems that are likewise strictly determined by material or efficient causes make sense for employing a machine metaphor. In the realm of medicine and human health, he uses this distinction to describe how the function, diagnosis, and treatment for the heart in many cases make sense to use the mechanical analogy—the heart can be thought of gainfully, and treated as if, a mechanical pump (albeit, linked to a larger circulatory system, kept functioning by energy of the digestive system, maintained by an endocrine system, and protected by a skeletal systems, etc.). Cancer or immune system disorders, however, cases with systemic causes and factors spanning more if not the whole of the human Life system, do not fit a mechanical analogy. For these more complex realms, the formal cause (and perhaps final causes) of Aristotle comes into play, as well as flexibility, contingency, and indeterminacy, and render the machine metaphor ineffective and even dangerous.

In a fun, scholarly, and highly informative debate on the pages of the journal, *Estuaries*, Bob Ulanowicz, and Bernie Patten debated the relative merits and flaws in the mechanical and deterministic worldview. One of many great parts of this dialogue came from Bernie's use of the case of flying an airplane, knowing that Bob is a pilot (as was Bernie), to evoke gut instincts to help compare the alternatives of mechanical and nonmechanical perspectives. Bernie (Patten, 1999) started it near the end of his review of Bob's book. Here, one needs to know that Bob employed Karl Popper's idea of propensity as a better fundamental and less deterministic/mechanical precept that Newtonian force:

Let me put it this way for my amicus who is a pilot. Sometime, perhaps near the middle or end of the next century, when some of the ascendency principles enunciated in his book have pruned...the world's airlines down to two giant international carriers, will you choose, at a premium price, the staid organization that gets you where you want to go the old fashioned way — *Newtonian Airlines*, or will you pick instead, at considerable savings, the cut-rate, flashier, and certainly more exciting carrier of the in-crowd — *Popperian Air*, whose planes have a well-known propensity to fly? (p. 342)

Bob replied (Ulanowicz, 1999b):

I have often joked about how I go from my office, where I rant against the idea of nature as machine, to the cockpit, where I pray I am seated in the most finely-tuned clockwork that ever existed! My appraisal of the two airways is different from Bernie's however: Newtonian Airways believes their machines are guaranteed by law to fly. Popperian Airways, chastened by the conviction that their machines only have a strong propensity to fly, invest in much redundancy so as to increase the probability that, when novel and threatening circumstances do arise, their planes will fail-safe. (p. 343)

This exchange is both entertaining and enlightening, and we learn at the end one major way in which Bob sees his ecological metaphysic as an improvement. By anticipating novel events and disturbances, he recommends a science paradigm able to inform the construction of needed and functional redundancy sufficient to allow adaptation and system reconfiguration to survive otherwise Life-threatening surprises. This redundancy, pervasive in the pathways of ecological networks, we have seen him document as critical, especially in the balancing act quantified by his robustness metric (see Chapter 6). His systemic measure of trade-off of these two key attributes—efficiency and redundancy—is one of the best single metrics for sustainability that we have encountered.

Sally Goerner provides another corroborating and expert view on the machine metaphor and the need for transformation in our ways of thinking. With advanced degrees in computer science, psychology, and nonlinear dynamics, she is another interdisciplinary scholar able to synthesize new ideas across multiple domains. We see her insights as particularly helpful as we seek to bridge between science and applications, between models in mind and realizations in the real world. Sally is also a masterful communicator, and her books, articles, words, and pictures have served to help communicate the most complex ideas in clear and readily accessible ways.

In her book, *After the Clockwork Universe* (Goerner, 1999), and in her other works, she has studied virtually all of human and cultural development and cosmic evolution to search for recurring patterns and general systems principles. As we assert in this book, she has targeted the paradigm, worldview, or shared cultural story as a fulcrum point on which to place a big lever for change. From her studies across multiple fields of human endeavor, she sees a "coming Great

Change," or Big Change, much like the shifts we propose as needed in science and culture. She wrote:

> We have built our society around clockwork beliefs, but these are simplistic and inadequate. We believe so strongly that our science is immune to Big Change, that we have a terrible time imagining how any other view could be valid — much less more powerful. We are like our ancestors. We need a new story, but we have a hard time seeing past the one which dominates now. This is why making the new science clear and relevant is so important. (p. 85)

And soon after:

> Beneath a calm, confident exterior, classical scientific images based on simple causality are crumbling. In their place a new science based on web dynamics (interdependence) and its organizing tendencies are emerging. In this completely understandable shift, all things change. (p. 85)

Here, we see again links between the machine metaphor with its simple causality on one hand, and what Goerner is calling the "web" metaphor, aligned with our holistic Life science, with its complex causality on the other hand.

After quoting John Muir, as we did above, Goerner (1999) went on to describe our "current" situation, which has spanned several centuries:

> In 1997 as in 1700 most educated people look out on a clockwork universe, a passive, essentially dead universe in which events are accounted for by mechanical forces. Throughout this century people have looked to scientific breakthroughs to end this view. Yet, despite many prophesies, none of the major scientific revolutions of this century — quantum mechanics, relativity, origins of the universe — have budged this core machine view. Some, like genetics, have tended to intensify it. Today more than ever the machine worldview seems unchallengeable and unchangeable.

> The machine view will end, however, because of a fatal flaw that is so simple that it is invisible...right now, the belief that things are fundamentally non-dependent (separate, separable and not connected) permeates our reality. (p. 85–86)

And a bit further on in the same book, she says this worldview will go away because it and its attendant specific beliefs, "are all based on assumptions of 'how the world works' in separate and disconnected pieces — and not together."

Goerner presents a concise pictorial model to contrast the two worldviews and their different approaches to causality. We reproduce her Figure 3 as our Fig. 7.3.

Referring to these diagrams, and to the way current world circumstances call for change, Goerner (1999) wrote:

> ...simplistic approaches to complex problems create dysfunctional answers and these dysfunctional answers are coming home to roost in societies the world over. This is creating pressure to find something better.

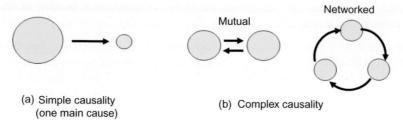

(a) Simple causality
(one main cause)

(b) Complex causality

FIGURE 7.3

Concise contrast of the machine metaphor (a), and its rigidly deterministic simple causality, "If A, then B—no exceptions!," as paraphrased by Ulanowicz (1997) vs. a nonmechanical metaphor Goerner calls "web dynamics" compatible with our principles of holism, complexity, and Life lessons, and the propensities and conditional probabilities of Ulanowicz, all gleaned from studies of ecosystems and networks. Redrawn from Goerner, S.J., 1999. After the Clockwork Universe: The Emerging Science and Culture of Integral Society. Triangle Center for Complex Systems. Chapel Hill, NC. USA. 452.pp. ISBN 9780863152900.

> ...it is increasingly obvious that most causality looks like Figure 3 [b]. In complex systems such as the body — not to mention economies, societies and the weather — causes blend and loop back on themselves. In such systems following single threads does not lead to *the* cause, but at best, a loop of causality and at worst, a hopeless confusion. (p. 91)

And in a passage that fits with our idea of the two-way bridge between science and the real world, and the dangers of unintentional overuse of the machine metaphor, she added:

> Scientists have known that the world is interwoven for a very long time. The problem is that the interwoven nature of the world can be overwhelming. It is overwhelming for scientific tools, social resources, and our merely mortal minds. The tools scientists use shape how they see the world. We all know the refrain: "if all you have is a hammer, then the whole world looks like a nail." Yet, researchers can only use the tools available...

> ...today we are stuck in perspectives that have passed their time. People's awareness of complex causality (webs) has been growing for years. A whole host of groups in every field imaginable have attempted to get other scientists and the public at large to realize that you can't understand much until you realize that everything really is a web — and that webs don't work like machines. Ecologists have been the most successful at this which is why people the world over are beginning to recycle and green movements are a force to be reckoned with. (p. 91)

Goerner's analogy of the available hammer tool influencing one to see nails everywhere is the same as one of the main ideas in this chapter—if you only have a single dominant scientific metaphor in mind, a mechanism or machine model, then the whole world and all problems or human needs appear as if they can be solved using a mechanistic approach. This leads to "dysfunctional" answers, solutions, technologies, policies, infrastructure, habits everywhere, and all the symptoms of systemic environmental degradation we depict in Fig. 1.2 are indeed "coming home to roost" as Goerner states. She also points to a pragmatic step for solution—we need more "tools" available. Thus, we propose several alternative metaphors and seek to qualify when the mechanical metaphor can be OK (as Ulanowicz noted for the human heart) and when nonmechanical models are fully necessary.

Frances Moore Lappé, who we have cited before for contributions to these topics, echoes the view of Goerner just above about the lead role of ecology in this transition of root metaphors. She wrote of the main theme of her book, which is about "changing the way we think to create the world we want" (Lappé, 2011):

> ...it starts with getting our thinking straight. Since we create the world according to ideas we hold, we have to ask ourselves whether the ideas we inherit and absorb through our cultures serve us. We can only have honest, effective hope if the frame through which we see is an accurate representation of how the world works. (p. 173)

Her focus on how we "create the world according to ideas we hold" matches the fabrication, realization, and technology application mode of the Rosen modeling relation (Fig. 7.2). Lappé and Rosen both are indicating how human ideas linked to our myriad action capacities can literally form a case of "mind over matter." Action and agency in this Anthropocene Era starts in our individual minds and collective intelligence and proceeds to transform the world.

Throughout her book, she shared her sense of hope that we can solve our systemic problems, but with the qualification that "honest hope" is not just naïve idealism or optimism. A bit later, in her final chapter that included her central concise recommendation for a solution to our interwoven social and environmental crises—"thinking like an ecosystem" (A title clearly in homage to Aldo Leopold's essay, "Thinking like a Mountain"), she wrote:

> Now we are realizing that ecology is not merely a particular field of science; it is a new way of understanding life that frees us from the failing mechanical worldview's assumptions of separateness and scarcity. (p. 174)

Lappé's emphasis on separateness here matches the focus of Goerner's critique of simple causality above and fits with our critique of analysis as an overused half of a full scientific process which requires synthesis for balance. Lappé goes into detail in her book on why ideas and fears of scarcity are connected to this mechanical worldview and linked shared cultural story. We highly recommend Goerner's and Lappé's books as essential reading for success in the human−environmental sustainability transformation.

Capra and Luisi (2014) provide the last set of corroborating ideas for the critique of the machine metaphor that we address here. In a comprehensive and valuable book that shared many of the goals and methods of this book, Capra and Luisi touch on Kuhn and the process of revolutionary scientific paradigm shifts, the moral responsibility of scientists in addition to the intellectual responsibility for their science, and many more profound topics. They chronicle and critique the ideas we have inherited from René Descartes. They acknowledge and appreciate many of the revolutionary insights and contributions of Descartes and his enormous impact on all the science that followed. They portray Descartes as a brilliant mathematician, an original thinker on par with Aristotle and Plato, a paradigm-shifting scientist and philosopher. The "crux of Descartes' method," as they saw it, is "radical doubt." We can see great value in continuing this practice, and we seek, as Descartes did, to radically doubt conventional knowledge and the current understanding of how the world works. "Radical doubt" fits comfortably alongside our founding holistic Life science principle of radical empiricism.

They also point to ideas they say Descartes got completely wrong, although admitting some errors were not known to be errors for hundreds of years. The first error they note is Descartes' "belief in the certainty of scientific knowledge" (Capra and Luisi, 2014). This, they say, has been corrected by the results of 20th-century science from Heisenberg to Plank to Gödel, which "has shown very clearly that there can be no absolute scientific truth, that all our concepts and theories are necessarily limited and approximate." Before this more recent scientific humility, Descartes believed that through exact mathematics, he could describe the world perfectly. As Capra and Luisi quote Descartes as boasting, "My entire physics is nothing other than geometry."

To describe both his strengths and how they can turn into problems, Capra and Luisi (2014) wrote:

> Descartes' method is analytic. It consists in breaking up thoughts and problems into pieces and arranging these in their logical order. This analytic method of reasoning is probably Descartes' greatest contribution to science. It has become an essential characteristic of modern scientific thought...and the realization of complex technological projects. It was Descartes' method that made it possible for NASA to put a man on the moon. On the other hand, overemphasis on the Cartesian method has led to the fragmentation that is characteristic of both our general thinking and our academic disciplines, and to the widespread attitude of reductionism in science... (p. 24)

Here, we hear again many of the same themes of our book and this chapter, including the dangers of unbalanced analysis. On the topic of the bridge between modes of thinking and "great" technological achievements, we must include both the great successes like getting to the moon, and great unintended technological side effects, like threatening the Life support capacity of Earth.

Descartes also played a key role in how we think about how we think, and how we imagine the mental and physical realms, as in the two main arenas of

Rosen's modeling relation—the formal system and the self, and the natural system of the world environment. Capra and Luisi wrote that from his famous starting point, that was the rock bottom he hit after taking his radical doubt to its extreme:

> Thus he arrives at his celebrated statement, "*Cogito, ergo sum*" ("I think, and therefore I exist"). From this Descartes deduces that the essence of human nature lies in thought, and that all things we conceive clearly and distinctly are true. (p. 24)

And soon after:

> Descartes' *cogito*, as it has come to be called, made mind more certain for him than matter and led him to the conclusion that the two were separate and fundamentally different. The Cartesian division between mind and matter has had a profound effect on Western thought. It has taught us to be aware of ourselves as isolated egos existing "inside" our bodies; it has led us to set a higher value on mental than manual work... (p. 24)

They go on to list many ramifications of this split of mind and matter, of mental from physical worlds, in physics, medicine, psychology, and even the marketing of products to seekers of an "ideal body." The dichotomy Descartes created as the basis of his massive program of scientific and mathematical analysis bears a striking resemblance to the modeling relation of Rosen. Capra and Luisi wrote:

> Descartes based his whole view of nature on this fundamental division between two independent and separate realms; that of mind, or *res cogitans* (the "thinking thing"), and that of matter, or *res extensa* (the "extended thing"). Both mind and matter were creations of God, who represented their common point of reference, being the source of the exact natural order and the light of reason that enable the human mind to recognize this order. (p. 24)

This division of realms and of topics of study, they tell us, led to the division between humanities and natural sciences in nearly all educational institutions that have followed.

Capra and Luisi also report Descartes' role in establishing the machine metaphor in science. They wrote:

> To Descartes the material universe was a machine and nothing but a machine. There was no purpose, life, or spirituality in matter. Nature worked according to mechanical laws, and everything in the material world could be explained in terms of the arrangement and movement of its parts. This mechanical picture of nature became the dominant paradigm of science in the period following Descartes. It guided all scientific observation and the formulation of all theories of natural phenomena until twentieth-century physics brought about a radical change. The whole elaboration of mechanistic science in the seventeenth, eighteenth, and nineteenth centuries, including Newton's grand synthesis, was

but the development of the Cartesian idea. Descartes gave scientific thought its general framework — the view of nature as a perfect machine, governed by exact mathematical laws. (p. 25)

They next describe how this shift ignited by Descartes came after an older worldview in which the world was seen more as a living thing than a machine. This new view of mechanistic science, they report, also led to changes in the value system and attitudes toward the natural environment. As they wrote:

The Cartesian view of the universe as a mechanical system provided a "scientific" sanction for the manipulation and exploitation of nature that became typical of modern civilization. (p. 25)

As we have noted, the interface from science to culture is a fast-acting portal, and Capra and Luisi documented how the scientific innovations of Descartes can be traced as actively transforming all of modern civilization and culture.

Capra and Luisi explain how Descartes also applied his machine metaphor to all living things. "Plants and animals were considered simply machines," and humans could be understood as having an "animal-machine" body "inhabited by a rational soul." As they and we have recognized, this metaphor can be appropriate and has yielded great achievements. The danger was in becoming too impressed with the success; as they described it:

The Cartesian approach has been very successful, especially in biology, but it has also limited the directions of scientific research. The problem has been that many scientists, encouraged by their success in treating living organisms as machines, tended to believe that they are *nothing but* machines. The adverse consequences of the reductionist fallacy have become especially apparent in medicine... (p. 26)

They go on to note that the limitations in this "fallacy" and shortcut of seeing living organisms, and other natural systems, as "nothing but machines" have been recognized in all the sciences.

This terminology from Capra and Luisi fits with our hypothesis in this chapter that the dominant machine metaphor in science has led to a kind of amnesia, perhaps ironically analogous to a fog induced by successive shocks to the collective head from observing just how successful this model can be. Imagine a proverbial hard palm slap to the forehead thousands of times—"It worked again!" Despite the jubilant successes (albeit ones that ignored full ramifications and externalities), we maintain with Capra and Luisi and the others we cite above, that it is a mental mistake with grave consequences if we apply the machine model everywhere and always while forgetting it is actually *nothing but a model*, and only one of multiple available models at that.

Capra, a physicist and systems theorist recently leading work in "ecoliteracy," and Luisi, a biochemist who has researched the origin of Life and self-organization of synthetic as well as natural systems, reach in their book many of the same conclusions and recommendations we make. Their book—another we add to the required reading list for the "coming Great Change" foreseen by Goerner—employs the metaphors and methods of systems as their central strategic approach. Their impressive and innovative book proceeds to apply the "systems view of life" to first biological realms, and then to cognitive and social fields, and includes a "conceptual framework that integrates these three dimensions" via systems thinking. In their final section of the book, they integrate "the ecological dimension in our synthesis of the systemic conception of life." Here and throughout the book, we see that their approach and ours are fully complementary and compatible. Late in the book, they echo our strategy and mindset (Capra and Luisi, 2014):

> The importance of studying, within the general framework of ecology, the pervasive influence of human activities on ecosystems, as well as the reciprocal influence of their deterioration on human health and well-being, also makes it clear that ecology, today, is not only a rich and fascinating area of study but is also highly relevant to assessing, and hopefully influencing, the future fate of humanity. One of the great challenges of our time is to build and nurture sustainable communities, and to do so we can learn many lessons from ecosystems, because ecosystems are, in fact, communities of plants, animals and microorganisms that have sustained life for billions of years. (p. 342)

Among many memorable and essential ideas in this passage, note especially that they suggest an option for another metaphor we can add to our "tool box" or "model repertoire" for understanding how the world works, and for guiding our interactions with it, and the technologies, applications, and behaviors we do in it. This alternative metaphor is the *community*. We return to this below as one of multiple options we propose can improve on and diversify our model repertoire toward the ultimate goal of sustaining and aiding Life itself.

But first, another bit of fun—our anachronistic reimagination of Descartes and the end point of his famous journey of radical doubt and inward reflection. Imagine how his insights, and the course of history, might have changed if ecology and the knowledge of ecosystems and the human place of interdependency in them were already well known at the time of Descartes' revolutionary work (somehow—perhaps if Priestley had come before and had used people instead of mice in the sealed chambers of his oxygen experiments!). Given this scenario of an altered knowledge and ecological awareness context, further imagine Descartes might have applied to thought our ideas about discrete Life and sustained Life. If in his radical doubting and precise thinking he focused more on continuous or "sustained thought" vs a short-term or "discrete thought" process, then he might have concluded:

> I think, therefore I exist. That is, I think now, therefore I exist now. But I think *only as long as* I have a steady input of oxygen, water, and food to sustain my thinking via my life. And, after I am done thinking with aid of these vital materials, they are transformed and expelled not so much as waste but as food for the plants and other living beings that in turn create and supply my material needs.

I think only as long as I exist in concert with the existence of these other life forms, and vice versa.

I think, therefore, I am. . .we are. . .an ecosystem.

We could play for hours with this rewriting of history, but we move on after noting how it helps highlight the core ideas in this chapter. With the actual historical timeline that occurred, including deep old ideas fully operative today of the separateness and fundamental difference of the mental and material realms, we have seen, observed, and have mounting evidence of the actuality that the machine metaphor threatens Life and by extension the continued thinking of Life's great human inventors, thinkers, and technologists. Since we have no mode of time travel and cannot change the Cartesian fork in history, we are left to heal this rift now and quickly remedy the situation.

Models, and ideas in general, while abstract, mutable, and unknown as to full details of their existence and workings, have very real impacts on the world. And, based on the necessity of a living human to create and use models and ideas, we also know that models can help or hinder the survival of those who think and use those models, and thus the survival of themselves as models. Models themselves can, in a sense, live or die; they can be "fit" and effective and thus continue to exist, or they can go extinct due to an inability to adapt to their environment, and to evolve new forms of their own model kind. If making the world into the machine image in mind is not working, or if the negative social and environmental side effects have surpassed the positive benefits, then fortunately we have powers of creativity and imagination to change our minds. This in turn can change the world we live in and co-create, and the course of history going forward. In fact, we assert a change in our mental model is the only way forward to reach our goal of organizing around sustained Life, since a single machine model is deeply incompatible with this goal. Furthermore, this change can be more immediate than searching for an as yet unidentified technological or political fix to right the ship to sustainability. We can begin today as we have the tools and knowledge to do so, and following in the shadow of that transformation, the ancillary issues will be easier to achieve. In the next section, we address some ways that transition has begun and other suggestions to further its appearance.

OUR SYNTHESIS IDEA—WE ARE MAKING THE WORLD INTO A MACHINE

It makes for longer reading and more time commitment, but we hope the direct quotes of the corroborating (and some dissenting) thinkers above is worth it—we see beauty and power in hearing the ideas expressed in the actual words of these allies themselves. As a combined concert of voices, we hear the chorus to ring loud and true—we have taken a good thing too far, we have mistaken our favorite model for the thing itself.

We think the growing worldwide evidence, and supporting insights from diverse unique observers, fit with our summary story:

> Based on an unbalanced, monolithic, dominant root metaphor at the foundation of the mainstream science paradigm, which prioritizes analysis over synthesis and reductionism over holism, we are accidentally manifesting that model, and, via the fabrication channel of Rosen's bridge, transforming the world through our mechanism-like technologies and industrial cultures, breaking the cycles of nature and fragmenting humans into an adversarial role with each other and the environment. We now realize that the world is breaking down, suffering stress and strain, and grinding to a halt like the mechanism we have unconsciously treated it to be for hundreds of years.

To grasp this full story requires many of the plot elements we have presented as in gestalt—the devaluation of Life, the fragmentation of Life from environment, the unbalanced overuse of analysis, externalization of environmental costs, and many more. The solution, then, likewise will require coordinated changes on many fronts—really a system of solutions spanning science, technology, and culture rather than any single solution in any one realm.

Near the end of his book, *Essays on Life Itself*, Rosen (2000) presented ideas on technology and "craft" linked to the fabrication half cycle of his modeling relation. In one section, he uses the example of a thermostat to illustrate how application of the machine model by itself leads to a slippery slope—technology that potentially can create more problems than it solves.

In his vignette, he describes adding a thermostat to the heating/cooling system of a room. This provides the added function of a control signal and loop as a "kind of dynamic insulator that *closes* our room from the effects of ambient temperature" (Rosen, 2000). The common human goal and intention is to create an intelligent device to regulate automatically the temperature of the room despite change in the temperature outdoors. This situation calls up the general incentives for convenience and comfort, for creating machines to do mundane and repetitive work, thus potentially freeing up the time of people for more creative, rewarding, fun, and enriching experiences, all of which are fine as starting points.

But, Rosen cautions and notes that adding the control loop—the material device of the thermostat and attendant parts in the temperature regulation system—also adds a new source of noise to the system of the room. This new noise, which also acts like friction to add an entropic and dissipative channel to the system, did not exist and was not an issue prior to the construction of the thermostat loop. Rosen comments on how the thermostat "closed," insulated, or buffered the room and humans occupying it to the noise and friction of ambient temperature fluctuations but at the same time *opened* the system to noise and friction impinging on the thermostat and its system and parts, creating new relations and dependencies on them. His example considers corrosion of the thermostat due to humidity and oxygen in the room.

Pushing this parable further, he observes that we could add another control loop to correct for the new source of noise, error, and friction, but quickly summarizes that this process as a whole leads to an infinite regress—such a technological strategy would have to add control loops forever. We propose that Rosen's short fable of the infinite regress and vicious cycle in this small example of the thermostat scales up globally to help us understand the world we now see. The layer upon layer of added control loop complexity all come with costs. Vast energies are devoted to managing control loops far divorced from and perhaps unaware of the original intent, leading to the proverbial situation of the "tail wagging the dog," taking us farther and farther from a sustained solution that exhibits process integration and closure to efficient cause. To understand another key facet of this situation, and how living organisms are similar yet different than the machines in terms of their entropy production, we return to Hornborg.

Hornborg (2001) also went back to Descartes in his work to understand and help solve the sustainability crisis and its connections to the machine. He noted that Descartes' machine model started as a model for a living organism. As the story goes, Descartes saw organisms work and appear much like automata, playful machines of his day, made to act in lifelike ways. Hornborg also addressed another angle on this landscape of many crisscrossing metaphors. Writing about how the metaphor of growth is confused between living systems and economics, Hornborg (2001) gives his own answer to What is Life? which is an excellent one that helps with the context we need for Rosen's infinite regress of mechanical control loops:

> To clarify how organic and economic growth differ, we must consider by which means these two kinds of "orderliness" (structure, organization) incorporate negentropy from their environments. For organic growth, the point of departure is the highly organized flow of energy that reaches Earth in the form of solar radiation. Life is the process by which the negentropy of sunlight further "informs" and animates Earth's thin surface layer of congealed matter-as-informed-energy. As the sun winds down by reconverting its own stock of matter-as-informed-energy into radiation, a very small fraction of this radiation transmitted in all directions is received by Earth and temporarily reconverted into structure being refracted into space in the degraded form of heat. This structure is the biosphere, a momentary, whirlpool-like by-product of the irreversible dissipation of the sun. (p. 123)

And a bit later:

> Because we can consider the input of sunlight available to the biosphere as a practically unlimited starting point, the closest thing to genuine "production" is photosynthesis, and plants are appropriately called "primary producers." From this point, each human act of energy conversion (from pasture and other crops through meat, human labor and technology to manufactured products) entails a net degradation of negentropy. (p. 123)

Negentropy may be a heady thermodynamic term for most, and the engineer's usage of the concept of similar term exergy doesn't help, but it becomes clearer when we think of it as work energy capacity. While energy is the ability to do work, the work done "degrades" its ability to do further work. Thus, while total energy is conserved (first law), its capacity to do further work is not conserved and in fact diminishes (second law), as evident in a decrease in exergy. Viewed on one extreme, this is the origin of the dreaded heat death scenario, but that angle omits the beauty and complexity of the Life—environment organization that arises out the work that is done.

Hornborg contrasts this context for the operation of Life forms with machines and industrial technology. His uses two facts to conclude that the "growth," "production," and overall effects are categorically different for Life compared to machines.

We had similar conclusions in our *Flourishing within Limits to Growth: following nature's way* (Jørgensen et al., 2015). The book took an approach that limits exist in natural systems but are not impairments to their success, complexity, diversity, or sustainability, etc. Human systems can learn to accept and flourish within these limits by recognizing and incorporating management strategies that consider 14 recognized ecological principles that deal mostly with energy, material, and information flows and organization. We compared and contrasted how and when socioeconomic systems are like ecological ones, with the following observation:

First, machines and industrial technology exploit and degrade nonrenewable and local (Earth) resources rather than renewable sources like sunlight arriving from outside Earth. Second, machines and industrial technology "feed on distant ecosystems by means of world trade" and thus are not constrained by local (community, ecosystem) negative feedbacks that check growth of a species population as its carrying capacity is reduced.

Hornborg, as well as perhaps all biologists and ecologists, recognize that living organisms are entropic in the same basic way as machines when single units of either are considered in isolation. Both depend on inputs of higher quality energy, both transform energy and materials to do work, and both export degraded and lower quality energy (among other exports). Thus, both organisms and machines do increase the entropy of their environments as they live or operate, respectively. However, the comparative similarity stops there, and Hornborg's two factors above reveal the essential distinctions.

Since the dawn of the age of Descartes and machines, we have fabricated many units, the machine has been replicated and realized billions of times, with each unit providing a useful function, but at the expense of increasing the entropy and degrading the quality of the environment. We assert this is directly correlated to degradation of planetary environment and the dozens of symptoms we depict in Fig. 1.2.

In stark and telling contrast, we do not see this same global impact with organisms. Billions and trillions of life forms have been produced, and they have existed and performed their individually entropic biological life functions. However, taken as a whole, and scaled up to the global spatial extent and full time history, we see just the opposite ultimate impact on the environment—all these organisms, integrated with their ecosystemic and biospheric super-systems, acted so as to *improve* the quality of the environment.

This nonentropic environmental impact is due to aspects Hornborg noted, and the many principles we have assembled in this book—organisms and Life are different because their primary energy source is both renewable and nonlocal to Earth; the unique Life capacity of improving material organization over time and via ecosystem and food web networks; material cycles and circulation aided by the biosphere; coupled complementary processes of autotrophic and heterotrophic forms; autocatalytic loops and their positive feedback and motive force; network mutualism and synergism; and the capacity for sustained life. Embodying and existing in all these ways, Life works to improve the local and planetary environment in syntropic fashion. It is true that highly degraded waste heat energy is emitted, but overall, Life units—in unified relation to Life itself—do work that alters the planetary environment for the better, so that the environment can do still more work to aid Life.

If we confuse the machine model for the world itself, and if we use living and environmental resources inside technological systems that are built on the machine model, then we can get more harm from the industrial systems than the living components are able to repair. Here are four brief examples of the repeated pattern of a short-term gain (like room temperature control) at the expense of longer term costs (like noise and entropic degradation added from the control loop and the incipient infinite regress of perpetual new problems):

Industrial agriculture—The "green revolution," gas powered mechanical plows and harvesters, genetically modified crops, nitrogen and other fertilizers, and many other machine advances have increased the amount of food we can produce in any given year on any given hectare. But this comes at the expense of degrading the necessary environmental basis for food production in the future—stable climate, healthy soils, species diversity, pollinator insects, and more.

Industrial food system—We likely now have more foods of more varieties available in more places than ever before. These are more convenient and less expensive (when considering money and market-assigned food costs as they currently exist, which are dysfunctional for communicating anything close to a "true cost" considered holistically). Yet, by many holistic studies, we are burning about 10 units of fossil fuel energy for every one unit of dietary energy delivered to a person. We also have huge problems with diet-related diseases and health degradation in industrial cultures like the United States In many places, obesity is a problem as expensive as, and causing harm similarly as negative as, hunger.

Industrial medicine—Using the machine metaphor beyond those cases that are appropriate, mechanistic medicine and its linked pharmaceutical industry have added an infinite series of treatments and pills (like control loops trying to regulate health) while the noise, system degradation, costs, and attempts at still more silver bullet micro-solutions all spiral out of control. This arena of mechanistic science and technology is also paradoxical. The whole enterprise treats individual human organisms as if they are "nothing but machines," working to swap out damaged or worn out parts or fluids, and to graft in new and technologically improved parts, but it also seeks to extend the lifespan and perhaps ultimately defeat death entirely. In so doing, this collective enterprise at times seems to have forgotten that all organisms die and all machines wear out and break down.

Industrial education and media—While the internet has been heralded as the backbone of the new Age of Information, and a new chapter in the history of modern education, during its recent explosion in use we have seen an erosion in understanding, civility, and democracy. Many universities provide massive online courses for free; huge collections of knowledge are available to answer questions in an instant, and world news is updated by the second, instantly, everywhere. And yet, we would be hard-pressed to say that knowledge, intelligence, or wisdom have actually increased. We are learning now that the brute force work of moving information around does not necessarily link to human mental, emotional, and spiritual processes of making sense of it and using it for the good of self and community.

If these examples show the unintended consequences of a single dominant machine metaphor in science, technology, and culture and also reveal the fluid conduit by which the machine metaphor *as idea* has power to transform the physical, material world and threaten all Life forms, then what strategy can work toward a holistic system of solutions able to affirm Life value and maintain Life itself and its environment context for being?

One of the best strategies we have sought to employ, and now recommend for the challenge at hand, is *dialectical thinking* as characterized by Peter Elbow. Working in the humanities, engaged with teaching literature and writing, and understanding learning, Elbow's brilliant approach employs models in a constructivist framework. In the passages below, Elbow (1986) grappled with the "structural difficulty inherent in knowing," which we see to apply to our current world situation. Not only have our own best modern ideas seemingly backfired and turned against us, but also we are in the predicament of trying to use a science still locked in the mechanistic worldview to diagnose, understand, and prescribe remedies for its own self-caused afflictions. This extended quote by Elbow (1986) addresses related issues. Just after addressing how observers can alter their own experiments, and confusions about objective vs subjective knowledge, he wrote:

> This epistemological dilemma has shown up particularly vividly in particle physics. Physicists cannot get information about a particle alone. They can only get a package of information about the interaction of the particle and the "observer" (i.e., the equipment). They can know the velocity of a particle, but not its location, or its location but not its velocity; but they cannot know both.

> The dialectical pattern of thinking provides some relief from this structural dif-
> ficulty inherent in knowing. Since perception and cognition are processes in
> which the organism "constructs" what it sees or thinks according to models
> already there, the organism tends to throw away or distort material that does
> not fit this model. The surest way to get hold of what your present frame blinds
> you to is to try to adopt the opposite frame, that is, to reverse your model. A
> person who can live with contradiction and exploit it — who can use conflict-
> ing models — can simply see and think *more*. (p. 241)

Elbow here provides a view of a new more self-conscious access to the model
we use and tells the reward for employing at least one additional model to our
most cherished one—this more flexible use of models allows us to see, think, and
know more. He went on to describe even greater benefits—this same dialectic
thinking can assist with revolutionary shifts, transformation in systems of ideas
and structured ways of thinking. Two pages later, he wrote:

> Searching for contradiction and affirming both sides can allow you to find both
> the limitations of the system in which you are working and a way to break out
> of it. If you find contradictions and try too quickly to get rid of them, you are
> only neatening up, even strengthening, the system you are in. To actually get
> beyond that system you need to find the deepest contradictions and, instead of
> trying to reconcile them, heighten them by affirming both sides. And if you
> can nurture the contradictions cleverly enough, you can be led to a new system
> with a wider frame of reference, one that includes the two elements which
> were felt as contradictory in the old frame of reference. (p. 243)

He continues with a description of how this breaking out of one system to find a
new and wider frame of reference fits with the way Einstein transformed the inherited classical science based on Newtonian mechanics. Elbow wrote about an early
example of this method of "embracing contraries" in Geoffrey Chaucer's work. In
The Knight's Tale, "Chaucer uses contradiction . . . to uncover the limitations of the
system in which he is working (chivalry) . . ." and ". . . to suggest a new, larger system" of values. This new larger system expands into a broader set of capacities
beyond chivalrous "courage, loyalty, honor" to include "feeling for others, humor,
irony, forgiveness, the ability to change one's mind, and the ability to grow and
change through suffering instead of just socially enduring it" (Elbow, 1986).

We borrowed this dialectical method to get to the ideas in Chapter 1—rather
than "neatening up" the current cultural and academic system that pits those
focused on sustainability against those focused on growth and development, we
sought to amplify the contradictions and find a wider frame in which the two
schools of thought, previously seen as contradictory, are both valid and potentially complementary. This general pattern also aided our work on discrete vs sustained Life, and it matches the thesis, antithesis, synthesis approach of Hegel.

In the introduction to his book, *Embracing Contraries*, Elbow (1986) wrote of
aspects of his quest to understand fundamental dynamics in teaching and learning.

Introducing the book, and describing what he had learned about his own development along the way, he wrote:

> A hunger for coherence; yet a hunger also to be true to the natural incoherence of experience. This dilemma has led me more often than I had realized to work things out in terms of contraries: to gravitate toward oppositions and even to exaggerate differences — while also tending to notice how both sides of the opposition might somehow be right. My instinct has thus made me seek ways to avoid the limitations of the single point of view. And it has led me to the commonsense view that surely there cannot be only *one* right way to learn and teach: looking around us we see too many diverse forms of success. Yet, surely, the issues cannot be hopelessly relative: there must be *principles* that we must satisfy to produce good learning and teaching — however diverse the ways in which people satisfy them. (p. x)

Following Elbow's inspiration, we move next not to a single alternative root metaphor for science, and not to a single prescription for how to bridge holistic Life science to a holistic Life technology able to achieve human–environmental sustainability. Instead, we present a multitude of alternative metaphors for the world and its workings which we hope may help stimulate "many diverse forms of success" in new ways of modeling and developing technologies and applications.

FROM MONO-METAPHOR TO POLY-METAPHORS: BLOSSOMING DIVERSITY IN THE TOOLKIT FOR LIFE SCIENCE

As we were drafting this chapter, one of us (Fiscus) had a father in heart surgery. The machine metaphor and the science around it, the vast enterprise that has grown and developed for centuries, has led today, among scores of other successes globally, to an 82-year-old man receiving the aortic heart valve made from cow heart tissue combined with mechanical parts to replace his own failing valve. The surgery was successful, and hopes are that Wilbur Guy Fiscus will benefit from many more years of life, and better quality of life unhindered by limited physical capacity, shortness of breath, and occasional heart failure in times of illness. In very personal, as well as scientific and philosophical aspects, we are not interested to discard the machine metaphor or mechanistic science as a whole. We do hope to suggest pathways for future innovation so that the gains accrue not only to fortunate individuals and their families, but also to all humanity, all Life, and all corners of our entire planetary home. While the machine metaphor works well for mending a human heart that is similar to a pump, we must honestly admit that this same heroic metaphor is terribly bad for addressing a broken heart from the complexities of human relations—since the control levers are so multitude and diffuse—let alone its ability to apply to Life as a whole including the intricacy of the biosphere with all its sentient beings, communities of diverse Life forms, interwoven webs of relations, and branching networks and loops of interdependencies.

As we propose a range of metaphors—like lenses of varying foci, or colored wavelengths of light that coalesce into what our everyday eyes see as visible white light—we build on all the principles of holistic science (Chapter 5), all the Life lessons (Chapter 6), and other core concepts we have covered. We don't need to review many of those here, but we do mention two topics again.

Given that we seek alternatives to the machine, a key property of the holistic Life metaphors we suggest relates to the internalism that von Forster et al. employed in their quest to develop a bridge for understanding nonhuman, very alien, but potentially autonomous beings in the cybernetic and artificial intelligences they sought to create and with which they sought to build pragmatic relationships. This same internalist approach was promoted by Salthe et al. toward better understanding of complex, hierarchical living systems.

Internalism helps to correct a profound liability of the machine metaphor that relates to the distinction between instrumental value and intrinsic value. A machine, such as a hammer, car, rocket ship, or other tool, has instrumental value—it is primarily a thing used to achieve some other end or purpose. This assumption of instrumental value is appropriate for true machines and tools, but highly problematic in nearly all other cases. All of the metaphors we suggest next are the opposite—these modes permit a view to see living things and the world not as instruments to achieve some other purpose, but primarily as beings with value in and of themselves. By extension, the task becomes not a project to use the world or any "other" to our own selfish ends, but to relate to and coexist with the world as a valued and respected "other," ambience, fellow community member, ally, and/or home.

The second topic we recall to guide the bridge between science and technology is the dichotomy between Sustainers and Transcenders. Since we have described these as fundamentally different worldviews and approaches to the current human−environment crisis, we recommend an initial step to clarify and choose consciously the worldview and associated cultural program within which one plans to operate. The crux difference, again, is that Transcenders may accept the idea of environmental limits, but their response is to transcend those limits, to innovate, grow, develop and use human technology, ingenuity, industry, and all capacities to alter the environment in any ways needed to allow continued human growth, development, increase, and expansion. Sustainers interpret signals and act categorically differently. They accept the idea of environmental limits, but their response is to live within those limits and to focus solutions and efforts for change on finding a "prosperous way down." This phrase from Odum and Odum (2001) describes the descent or "soft landing" by which industrial culture reduces its energy use, materials extraction, waste emissions, global footprint, and consumptive ways while also increasing quality of life, freedom, social equality, and economic equity for people.

Some distinctions can help to describe the differences in the bridge between science and technology, and the technology and applications, for these two hypothetical camps. The Transcenders are likely to continue to need and benefit from the machine metaphor and mechanistic science and technology. The most obvious example of this is the space exploration, space travel, space stations, and eventual space colonization program that is ostensibly led by the Transcender faction. (Also, recall that while we have posed these factions as truly different and important to separate, we also pose them as fully interdependent, complementary, and both needed in Life as well as human Life.) But, other projects under the Transcender banner will likely continue to benefit from machines; ideally, these would be used mostly for those purposes and cases where the machine metaphor is most beneficial and with fewest negative side effects, like the human heart as pump scenario and many more. Continued use of machines—like a blend of trying to sustain the Transcender program—would require creativity, discovery, and substitution as materials and energy sources run out, and negative side effects accumulate. Or, perhaps conversion of machines to run on (and with manufacture based on) renewable energy and recycling materials processes will be adopted as a hybrid way to sustain the use of entropic machines as long as possible. As we have seen, continued use of the machine metaphor and mechanistic technology will also require attention to costs and repair of environmental degradation.

The new candidate root metaphors we describe next are primarily intended for use in the Sustainer program. This effort, as we have justified in this book, is now the appropriate main mode of action for Life on Earth. As such, the lenses and colorful diversity of approaches we propose—not just rose-colored glasses of naïve idealism but also a suite of radically empirical views to help us see and achieve pragmatic transformation—are nonmechanistic metaphors and ways of framing systems of study, research and development projects, technological applications, education, and more. Individually, and as a whole, these metaphors and the "bridge" they form in concert are inspired by Life—Life itself, Life as a unified whole, Life-and-environment in win—win and mutually beneficial relation.

If the Sustainers and Transcenders can achieve a synergistic peace accord, then we look forward to ultimate success, truly grand human and Life achievements, on both fronts. Life will be Sustained and the corner turned quickly so that environment and Life support systems begin immediately to heal and regeneration proceeds quickly. And, we continue to Transcend those appropriate and challenging limitations as inspired by the spirit of exploration, discovery, and creativity. Clarified thinking and full cooperation will be required, as continued fighting drains essential resources, delays progress, creates new unnecessary problems, and threatens both programs with failure. Discord thus threatens Life in both its Earth and beyond-Earth potential futures.

Rosen confronted the machine metaphor as limited by an "impoverishment of entailment," and he described how and why this makes questions like What is Life? impossible to answer using the mechanistic paradigm. We likewise confronted the sustainability crisis and problem of humans-in-the-environment and see the mechanistic paradigm unable to provide effective answers or solutions. Rosen also showed how relational modeling inspired by biology is better and has a richer and more general capacity to understand and describe entailment. His model for organismal Life had a nonmechanistic feature of closure to efficient cause, and he also described ways in which life forms are built in ways that differ from machine construction (see his example of a bird and bird wing compared to an airplane and airplane wing; Rosen, 1991, 2000). Aspects of complex causality and entailment including self-reference, impredicativity, ambiguity, and inclusion of efficient, formal, and final causes are all features in the metaphors we list next. Rosen also suggested that relational biology is more generic, more widely applicable than the mechanical paradigm, which he saw as only applicable to rare, constrained, special, and artificial cases. If this insight holds true, in addition to opening vistas and projects to achieve human−environmental sustainable, then the science and technology able to resolve the paradox of human action in the environment should lead to new leaps forward in science itself.

Our candidates for new root metaphors as guide to technology and application for holistic Life sustainability are:

1. Life—Organism (but not alone), ecosystem, biosphere, Life itself. This metaphor applies to any living system, and we urge use of all holistic Life science aspects as above in this book. As applied to physical and material subsets of the universe, this approach to use Life as metaphor can focus on Life support (atmosphere, oceans, hydrological cycle, geological processes, etc.). It can also be useful to understand and inform those capacities inherent in the physical-material universe that are lifelike or hold potential to aid, as well as threaten, Life. These include self-organization at multiple scales, interplay between solar (or stellar) radiation and planetary gravity, ways in which energy can be seen to relate to information, etc.

2. Network, web, or ecosystem—As we have seen, the network metaphor and model, as employed in ecological network analysis and network environ analysis, has yielded multiple insights and discoveries of how Life is organized, functions, and successfully achieved self-sustained operation over millions and billions of years. Goerner employs the web version of this metaphor often. And, many references to the ecosystem metaphor are used now—people speak of the "healthcare ecosystem," "media ecosystem" and many other cases where diverse participants interact and coexist.

3. Community—Closely aligned with the ecological metaphors just above, the community is useful in its accessibility in the sense of a human community. Many people can grasp how dynamics in a human community (which also apply to other Life communities) depend on a diversity of roles, complex webs of interdependency, resolution of conflicts, achievement of synergy, and many more nonmechanistic and fully necessary Life-affirming capacities and attributes. Ironically, the early colonial days in America were characterized by

the rugged individualist. And, while there is no doubt some of those individuals existed, the real success of the small frontier town rested on community and the functioning of roles and responsibilities in that ecosystem (in both the natural and organizational sense). Success of small town America is often, reductionistically, attributed to a lack of government regulations which meant that unencumbered free market principles controlled decisions. This revisionist history overlooks the role of the community in these communities. Each small town had one baker, one butcher, one candlestick maker, etc., because those were essential products for survival. Each was a monopoly, but prices were constrained by a community embrace not an invisible hand. In a truly local economy, what good does it do for the butcher to raise prices? First, there is no hiding in a community so that there are immediate personal condemnations, but even from an economic perspective, the price hikes return as the other businesses raise theirs in a pointless inflationary exercise. Community relations can be extended beyond people—people and people—group interactions to include the people—environment ones, such that local place and resources—plants, animals, streams, and hills—take on real meaning and relations.

4. Family—Yet another slight modification on 2 and 3, and again well suited to striking a chord of awareness with people. It has become more commonly known that all humans form a single family of related kin (or, in biological taxonomy, a species), all descended from the first *Homo sapiens*. This metaphor can extend beyond humans to Life as we see the same unifying kinship and relation as we continue to understand how all Life has descended from original Life forms. Another positive attribute of the family metaphor is that it helps with awareness that race is an artificial division with no meaningful implications for any differences in rights, treatment, or equality among unified family members.

5. Mind—This metaphor borrows from Gregory Bateson and others who focus on the inherent intelligence, or capacity for information and intelligence, in the natural world and the universe. This metaphor has grown via path-breaking work of scientists like Jane Goodall who helped us better get inside the mind of chimpanzees and our other great ape relatives. It may be useful in scaling up to imagine how individual minds, unique and diverse intelligence of many individual people, can combine to form collective intelligence even greater than the sum of the "parts" as in contributing minds. This positive outcome would help to counteract some aggregation problems we see now such that national governments may be more fearful and militaristic than the many compassionate, generous, and kind individuals in the country as a whole. The same might be said of paradoxical lack of intelligence, wisdom, and capacity for decision-making of the US government despite the myriad skills and mental prowess of its individual members.

6. System—This is the most abstract nonmechanistic metaphor, but very useful as Capra and Luisi, the fields of systems ecology and general systems, and

many more have shown. Systems are excellent in their flexibility and adaptability—their abstract nature allows them to fit to ecosystems, social systems, economies, corporations, cultures, cities, states, and many more organizations in need of modeling, understanding, and guidance for sustainability. We could include holons in this category, too. By design, this metaphor is holistic, looking for leverage points and anticipating unwanted consequences.

7. Sacred—Our only transcendental metaphor, we see potentially great value in using a metaphor based on ideas of a higher power when interacting with either living or nonliving subsets of reality. This metaphor would be useful by acknowledging mystery, the unknown, and those areas that may be inherently unknowable to humans. As with Schweitzer's reverence for Life ethic, this metaphor would involve approaching any system of study, research, development, or technology with humility and respect for that system as a creation of a supreme being (or at any rate, a creation of something other than humans). Clearly, this metaphor would be best if nondenominational, and thus it would also face challenges due to different religions and spiritual traditions. However, we see the potential benefits to outweigh the risks or downsides (albeit requiring great wisdom and tolerance). Examples of generic principles amenable to such modeling of "world as God" (or godlike) include creator/destroyer (or creation/destruction) aspects inherent in all systems. Panentheism is an interesting example approach that sees such normally godlike capacities in everything, or one could say unfractioned unity between God and the universe (Shani, 2014). Bob Ulanowicz has also written of ecology as a "natural middle" between spiritual and material approaches to understanding the world.

We also see these metaphors as useful to blend and mix as needed—they can be modified with multimetaphor, hierarchical combinations, nested arrangements, or other hybrid metaphors to suit studies or applications as needed. Any of the above nonmechanistic metaphors can be combined, and we can combine any of the above with the machine metaphor. We are also aware of others that could be added to this list—fractal, forest (as in the story, "The Word for World is Forest" by Ursula K. LeGuin), relation or relationship, lover, and self (as in the book, *World as Lover, World as Self* by Joanna Macy), and the Tao are a few more examples of models focused on things or entities. We could also add learning, inquiry, dialogue, conversation, and relationship as holistic nonmechanistic metaphors focused on change and process rather than object or objective.

SUMMARY AND STEPS TOWARD CHAPTER 8

Our preceding Chapters 3–6 were efforts to propose a holistic Life science, and Chapter 8, which comes next, is our outline of technology and applications able to truly solve the human—environmental sustainability problem. In this chapter, we have proposed a necessary holistic bridge to unify the science and the applications, like a means to integrate thought and action so that both are tuned individually and harmonious in concert to serve Life and benefit humanity. As with most other topics in this book, we have been informed and inspired by Life itself in all of this work.

While we could have included this example below in Chapter 4, on holistic concepts of Life's essential nature and origin, we can use it now to help summarize our current discussion. Alf Hornborg, though primarily an anthropologist and human ecologist as stated above, also provided an answer to What is Life? He wrote (Hornborg, 2001):

> Life is the process by which the negentropy of sunlight further "informs" and animates Earth's thin surface layer of congealed matter-as-informed-energy. (p. 123)

This is a useful short answer and it unifies the major principles of ecosystem ecology—energy flow and material cycling. It also fits with the complementarity of ecological goal functions, and with our quest in this chapter for a Life system with intelligence that integrates matter, energy, information, and self-sustaining behavior all in concert with the natural operating principles of the planetary and solar system environmental context.

We have mentioned before two additional models that match this ultimate objective of intelligent, holistic, unified thought-action, like unified science-technology. Odum (1971) wrote that his own scenario for an ecological origin of Life was an example of a "choice-loop-selector." Bateson also wrote of a "trial-and-error system" as linked to both a unit of survival (and Life, and evolution) and as a unit of mind (Grossinger, 1978). Bateson later (1988) contributed a list of criteria by which a system could be determined to qualify as a "mind", and thus, require treatment as such. His criteria overlap with many of our principles, and they include "interaction between parts of mind are triggered by difference," "mental process requires collateral energy," "mental process requires circular chains of determination," and more. He then wrote of his list:

> I shall argue that the phenomena which we call *thought, evolution, ecology, life, learning*, and the like occur only in systems that satisfy these criteria. (p. 98)

Hornborg, Odum, Bateson, and we envision the potential for humans to mimic this full integration of mind, learning, environmental context, and Life-affirming thought and action. These three independent views are mutually corroborating and they also help support our idea that Life's great intelligence and capacity includes the ability to *improve the environment* over time so that the *environment is better able to support Life*. This ultimate outcome of Life's integrated intelligence-action is clearly nonmechanistic—no machine is able to leave its environment in better condition. All machines export degraded energy and materials and leave the environment with greater entropy over time. Machines are very useful, but not for the systemic task we have at hand.

Another important topic relates to final cause. Some researchers and thinkers treat entropy production and the second law of thermodynamics as if these provide an appropriate final cause for Life and humans as well. While we accept that entropy increases in a closed system as required by the second law, we also see a concomitant self-organization and negentropic process always involved as well. Ulanowicz (personal communication) has written that while it is true that you cannot do work without dissipation, the inverse is also true—dissipation cannot occur without associated work being done and orderliness increasing. These two tendencies appear unfractionable and in dialectical and complementary coexistence. It thus seems possible that syntropy, negentropy, and self-organization could be equal to, and at times even greater than, entropy and dissipation.

We think the hope and optimism required, as well as the pragmatic science and technology for implementation of the solutions, call for some other ideas of final cause for Life and humanity beyond merely aiding entropy. To learn and evolve; to nurture, sustain, and develop Life on Earth; and to strive to extend Life beyond Earth—these are ultimate or final causes that go far beyond machines and the entropy they inevitably produce.

When we integrate our Life value and Transcender/Sustainer typology, we could build on Hornborg and say that Life is the process of harnessing sunlight so as to continually inform and improve matter to both sustain Life on Earth and work toward the potential to Transcend the Earth and colonize Life beyond Earth. Any industrial process, and any scientific, economic, or cultural system, that harms or threatens these core Life projects must be seen as ultimately detrimental to Life and humanity. Any proximate or currently accepted ideas of benefit (like profit, fame, or power) from such Life-degrading enterprise must be seen as delusional, dysfunctional, unjust, and in need of immediate remedy. A recently discussed idea, including a patent for which Walmart has filed, to create robotic micro-drones to do pollination services in place of bees that are now declining provides one of many examples that help us make our case (Hetherington, 2018).

Andy Clark, a philosopher and cognitive scientist, developed and has promoted the idea of the "extended mind" that is compatible with our vision of ecological intelligence that extends into human culture. His idea is based on his awareness that cell phones, calendars, tools and even people, books, and academic talks all can serve to allow us to think thoughts we could not think without these natural extensions of our minds. MacFarquhar (2018) reported on Clark's recent ideas and advancements including his status as "one of the most cited philosophers alive" whose ideas have influenced fields of neuroscience, psychology, linguistics, artificial intelligence, robotics, and more. The original 1995 paper, "The Extended Mind," he co-authored with David Chalmers. Just after publishing it, Clark "began thinking that the extended mind had ethical dimensions as well." (MacFarquhar, 2018). She elaborated on this aspect of Clark's work:

> If a person's thought was intimately linked to her surroundings, then destroying a person's surroundings could be as damaging and reprehensible as a bodily

attack. If certain kinds of thought required devices like paper and pens, then the kind of poverty that precluded them looked as debilitating as a brain lesion. Moreover, by emphasizing how thoroughly everyone was dependent on the structure of his or her world, it showed how disabled people who were dependent on things like ramps were no different from anybody else. Some theorists had argued that disability was often a feature less of a person than of a built environment that failed to take some needs into account; the extended-mind thesis showed how clearly this was so.

We see our current suite of systemic environmental problems as a similar form of "disability" in our built environments, thus also implicating the science and technology that helped construct them.

Clark also interestingly admitted that his extended mind theory isn't primarily a factual claim—"you can make a case either way." MacFarquhar (2018) reports, "No, it's more a way of thinking about what sort of creature a human is." Our holistic Life science and the nonmechanistic bridge to technology may also be more about "a way of thinking"—not really an argument, and with no absolute right or wrong decidability. But, we have presented multiple new metaphors from which to choose with evidence of their utility and help, different costs and benefits, and potential to yield different ultimate outcomes. Much like scenarios, these alternative ways of thinking can provide a structured and systematic way to consider the future and to help us in choosing positive futures and avoiding negative ones.

The choice and the alternative path it implies do not come without significant costs as well. Not only must we accept, adopt, and work with the death of individuals as normal and natural, but we must also become almost schizophrenic in the sense of having at least two senses of self. The first self is the current and traditional idea of the human self as bounded by our skin—an individual person as an instance of discrete Life, with a finite life span, with a certain death ahead, and who, like a machine, must succumb to the second law of thermodynamics. But, a second necessary self is the ecosystemic, biospheric, or "ecological self" (Arne Næss term) of sustained Life. By identifying with sustained Life, that need not decay or end, that can and has lived on for billions of years, we gain the power to make a new kind of choice based on the newly elevated value of Life. We can choose to serve Life and its improvement in quality and order even as this process leads to sacrifice and decay of our individual selves.

We next describe the technology and applications including the qualities in them we seek for sustainability, and we present case studies that demonstrate these qualities in action. As we continue to understand the Anthropocene Era in which humans now shape the planet more than geological forces, the stakes increase that we shape the planet for the better. Life, mind, and their extensions into human culture can in some real ways be more powerful than the physics and chemistry of matter. As we accept such awesome responsibility, successful planetary stewardship will require courageous choices, great creativity, and many more of the best traits that we have developed as humans.

Technology and applications in the context of holistic life—environment

INTRODUCTION

In this chapter, we look at technology and applications with break-through capacity to contribute solutions to the systemic human—environment problem. We focus on existing examples and case studies as these provide specific information and context by which to discuss successful instances of translating holistic Life science into real-world projects, systems, and creations. Not that we pretend to be science-fiction writers, but we also discuss several potential or hypothetical applications and technology beyond what exists now, knowing that history has shown often technologies are anticipated in such a manner.

Our first distinction is between (1) applications and technologies that explicitly use the value and philosophical foundations and holistic Life science we have presented versus (2) those that are compatible to varying degrees and successful with pushing human—environment relations toward win—win outcomes. Of course, there is a sundry of other technologies available that do not honor the Life—environment principles, many fall into the category of geo-engineering (e.g., seeding the oceans to promote algae growth for carbon sequestration, or release of stratospheric sulfur aerosols to induce global dimming)—all of which employ a machine metaphor without full accounting of unintended consequences. We do not address those technologies here since they are inconsistent with the aims of the Life—environment perspective.

The first category of applications and technologies *fully* founded and based on holistic Life science has very few examples (but see below for closely compatible examples). In making this distinction, we are strict in requiring near total overlap with our major conceptual framework, six principles, and seven Life lessons. In many ways, since this school of thought and paradigm is new, it is most likely to exist and operate in the realm of education or research rather applications and technology. The few examples of which we are aware include the continuing scientific and written works of Patten, Ulanowicz, and Goerner. The main organization we can cite is the Research Alliance for Regenerative Economics. Even these, our closest collaborators, allies and like-minded folks, and ourselves included, do not have, as well as utilize, several resources which we see as necessary—for example, new textbooks with a different conceptual approach to the

Foundations for Sustainability, DOI: https://doi.org/10.1016/B978-0-12-811460-5.00008-X

origin of Life and the answer to "What is Life?". Also needed are new textbooks, dictionary definitions, and references that begin the study of Life science with an understanding of the full integration of Life—environment and the essential win—win relation such that environmental quality improves over time. We hope that this book can serve as one starting point and, by integrating other holistic Life science works, can help with the processes of generating these necessary educational resources soon.

The second category includes options that share a very high degree of overlap with holistic Life—environment science in the concepts, values, metaphysic, science, and educational programs. The Center for Ecoliteracy and Schumacher College are two leading examples in this category.

This chapter will NOT be about "low-hanging fruit"—the common environmentalist approach of seeking primarily those changes that are easy, inexpensive, and require minimal effort, such as changing light bulbs, increasing fuel efficiency for cars, etc. Instead, we seek to focus on and work at the level of the "deep tangled root" (Fiscus, 2013) to transform the model, template, image, vision, and mission of technology itself, not merely to tweak it cleverly for profit, incremental improvement, or worse, a false "feel-good" sense of accomplishment. These are all examples of reducing unsustainability but not necessarily becoming truly sustainable. Furthermore, there is much evidence and many papers written about the rebound effect that such incremental improvements actually increase resource use and environmental degradation over the long run. Technologies that provide greater efficiencies are typically not used only to produce the same amount with fewer inputs but to produce more with the same inputs, until positive feedbacks reinforce even greater uses of the technology and resources. In fact, in a nutshell, this is the story of growth in civilization (particularly since the industrial/fossil fuel revolution). Our cause is against powerful forces, but our aim is not simply at minor tweaks that are just "neatening up" the system but rather breaking out of the paradox completely (Elbow, 1986). Another tree in the forest improves the overall health of the forest ecosystem by organically fitting into it Life—environment context. How can humans do the same?

While only partly overlapping in stated or explicit terms with our foundations in this book, many organizations, individuals, groups, and agencies are doing excellent work in applications and technology. The general principles, qualities, and unique features of a holistic Life-affirming technology include the following:

1. Unlike machines and technology based on the scientific machine metaphor, holistic Life technology will result in net improvement in the material and energetic capacity of the environment to support Life. While the machine metaphor can work at times, no subset of Life on Earth is actually machine, and thus nothing is simple for science or modeling. We must relate to and manage all Life—environment systems differently, and we must treat them as complex in the sense of Rosen and use multiple and nonmechanistic models as described in Chapter 7.

2. Similarly, holistic Life technology will lead to a net increase in the orderliness of the natural and built environment and thus tend toward increased syntropy and decreased entropy. This effect is achieved primarily by (1) operation based on renewable energy; (2) systems with a very high degree of material recycling; (3) mimicry of the other processes, principles, and holistic relationships integrating organisms, ecosystems, biosphere, and environment as we have described in this book and our prior book (Jørgensen et al., 2015); and (4) whole system network connections that result in process couplings. An observable change related to points 1 and 2 is that holistic Life technology will result in an increase in functional gradients in environmental systems just as we see Life systems to build functional gradients in vertical profiles in soils, concentrations of atmospheric gases, very high levels of diversity, and surplus energy as in fossil fuels.

3. Holistic Life technology is anticipatory and serves long-term goals and ultimate Life value by protecting Life and its essential environmental context. In the near term, it will also enable and catalyze a transition during which past environmental damage is restored and regenerated.

4. Holistic Life technology is self-referential and uses an internalist orientation to account for its own impacts on the environment, to conserve and plan for its own needed inputs and waste emitting capacities, and to treat other living systems and the environment as having inherent value and autonomy. Thus, holistic Life technology not only helps sustain Life but also sustains itself, and the science that supports it. Internalism is important, too, in that we understand that we are not outside (not really objective), and we are not able to control any system of study when that system is autonomous. This new orientation requires new ideas, methods, applications, and technologies that get beyond the ideas and goals of control, having power over, and treating Life—environment as purely of instrumental value. This can include cooperation, synergy, biomimicry, mutualism, interdependence, community, coevolution, win—win relations, and more.

The applications and technology we propose, and for which we discuss existing case studies below, also will integrate the six principles of holistic Life science and the seven Life lessons in the supporting holistic science that develops and serves to refine, modify, upgrade, and evolve the technologies going forward.

While technology that reverses the current trend of damage and degradation to the environment of existing industrial technology is an ideal, we have shown that it is not naively idealistic, romanticized, or unobtainable. In Chapter 3, we have shown solid evidence that a net result of environmental improvement over time is not only possible but is also the norm for Life systems. Self-organizing, self-sustaining, and self-enhancing impact on atmosphere, soils, biodiversity, and those deposits that became fossil fuels—all of these ***prove*** that it is fully and practically possible to operate a large scale, complex Life system while continually preserving and enhancing the Life support capacity of the planet.

In Chapter 7, we also presented an explanatory narrative that what we see now—the degradation of atmosphere and soils, the destruction of biodiversity, the depletion of fossil fuels—symptoms analogous to "breaking down," "wearing out," and "running of out gas"—are *symptoms that apply primarily to machines.* As such, these symptoms make most sense as observations of the living world when we realize they have been generated by humans, by our single-minded application of the machine metaphor in science and technology. This creates a vicious cycle in that we observe outcomes emanating from the worldview which to now only reinforces the worldview to double down on machine metaphor solutions. These symptoms of systemic decay we see now make no sense as observations of healthy Life—environment systems, which naturally do the opposite.

The fact that the closely intertwined scientific, technological, cultural, and environmental realms all mold each other, while problematic in the recent past and present, can be of benefit in the future. By transforming science as with the new foundations in this book, and transforming the bridges to technology and culture, we can begin the systemic change, Great Change, and Great Transition processes we need to eventually heal our relation to the environment. This will depend on science and technology in which we no longer have such extreme imbalance between analysis and synthesis and no longer treat the world as a machine. These changes will allow our planetary home to restore its own natural Life processes including the essential win—win Life—environment relationship. If fully developed and tested, then this holistic Life technology we propose has the added benefit of making it possible to develop sustainable and self-enhancing Life support systems off-Earth—during space travel, on space stations, and for colonizing other planets.

HOLISTIC LIFE TECHNOLOGIES DIFFER FOR SUSTAINERS AND TRANSCENDERS

In addition to the major contextual conditions above (need for science paradigm reform, pragmatic possibility of win—win environmental relation, internalism, understanding the machine metaphor, etc.), we can also benefit from zooming out in time to consider the very unique quality of this moment in history from an energy perspective. With energy, as with everything about the planetary and species-level "phase transition" we are moving through, we need to clarify the self-world (and human—environment) context for two seemingly opposite yet necessarily complementary worldviews, the Transcenders and Sustainers.

In Fig. 8.1, M. King Hubbert in 1976 (citing his own prior work, Hubbert, 1962) depicted the fossil fuel era as a very dramatic and very brief event in human history when considering a long-term perspective. His choice of 10,000 years is an interesting time frame, as this is about the length of the Holocene, the period of stable climate during which most of human agriculture, technology, and cultural development has occurred. Hubbert has shifted the time frame, however,

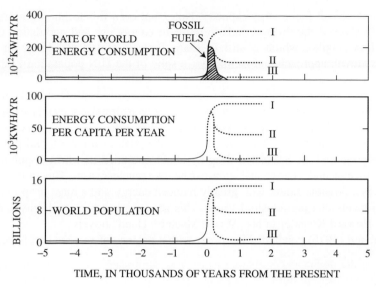

FIGURE 8.1

Key variables of energy consumption and world population over 10,000 years and under three future scenarios.

A Report to the Committee on Natural Resources of the National Academy of Sciences-National Research Council (1962)

to look at a period in which the fossil fuel era is in the middle of the 10,000-year period.

Hubbert, a petroleum geologist who worked for Shell Oil in Texas, became famous for correctly predicting the peak in US domestic oil production long before it actually occurred. He was doubted, dismissed, and ridiculed for years until the data showed he was correct, if off by a year or two in the timing of the US production peak. (Note, peak oil production concerns are currently out of vogue given the recent rise in domestic supply from fracking, but this unconventional source does not refute the continuing decline of conventional sourced production. Also, fracking, tar sands, and other unconventional supplies are nonrenewable, cause greater environmental damage and some unknown risks, and yield less net energy than conventional supplies. These supplies do not alter Hubbert's 10,000 years curve and provide no hope or steps toward solution. They are another example of kicking the can down the road and denial of the inevitable.)

Hubbert's three scenarios in Fig. 8.1 (labeled I, II, and III) for events after the correlated spikes in energy consumption, energy consumption per capita, and global population, had to do with possible futures for energy, technology, and human population after the fossil fuels are gone. In his original report (Hubbert, 1962), he explained the high energy, high population scenario (I) as a case in which nuclear power and other energy sources were harnessed to fully match and replace the fossil fuel energy capacity. Note, that even this techno-optimist future scenario showed stabilizing trends in terms of energy use and population (around 14 billion people), which is still in sharp contrast of today's dominant never-ending growth approach and mentality (in spite of the U.N population projections that anticipate a leveling off between 10-12 Billion; it contradicts the growth imperative of debt-based economy; and no evidence exists of any country willingly and proudly aiming for stable population as national policy). His intermediate scenario (II) was a case in which nuclear and other energy sources are only able to replace about one-half the energy of the peak in oil, coal, and gas energy consumption and an associated drop by about one-third in world population due to "confusion and chaos" and catastrophe like nuclear war. The third scenario (III) was a possible future with greatly reduced energy and a huge drop in population to levels like preindustrial times. This scenario has been well envisioned by James Howard Kunstler in his "World Made by Hand" novels.

Hubbert (1976) concluded his paper with a balance of realism and hope:

> Since the problems confronting us are not intrinsically insoluble, it behooves us, while there is still time, to begin a serious examination of the nature of our cultural constraints and of the cultural adjustments necessary to permit us to deal effectively with the problems rapidly arising. Provided this can be done before unmanageable crises arise, there is promise that we could be on the threshold of achieving one of the greatest intellectual and cultural advances in human history. (p. 84)

Today, more than 40 years later, we are still in a similar place with respect to the need to "deal effectively with the problems rapidly arising." Just before the passage above, Hubbert had written:

> Our principal impediments at present are neither lack of energy or material resources nor of essential physical and biological knowledge. Our principal constraints are cultural. During the last two centuries we have known nothing but exponential growth and in parallel we have evolved what amounts to an exponential-growth culture, a culture so heavily dependent upon the continuance of exponential growth for its stability that it is incapable of reckoning with problems of nongrowth. (p. 84)

This too fits with our current circumstance, and our view (corroborated by Lappé, Meadows, Goerner, and many others) that the cultural mindset, root metaphor, mental model, paradigm, value system, and links between science and society are critical for change.

However, relative to Hubbert, we think our framing of the two worldviews, Sustainers and Transcenders, plus the other major principles of this book, provide a different and even more hopeful perspective on the possible scenarios. We push Hubbert's optimism further and believe firmly *that we are indeed* "on the threshold of achieving one of the greatest intellectual and cultural advances in human history" (albeit decades later than he may have hoped).

Without considering that Sustainers and Transcenders are truly and categorically different, that both are needed, and that neither one is likely to "win," eliminate, overpower, or convert the other, we would be left with scenarios only related to Life on Earth, and assuming futures within other existing human and Life—environment constraints, like Hubbert's three above. But given our framing of these two cultural types, on top of advances in space travel and technology and other novel innovations since 1976, we can add in the issue of transcending environmental, spatial, and human limits on Earth. We can potentially integrate the human cultural projects of transcending environmental and human limits. Such projects include the main example we have used, colonizing Life beyond Earth, but could potentially include others such as transhumanism or genetic engineering. Such high-tech and mechanistic endeavors may seem at odds with reverence for Life, but in the effort to embrace even antithetical opposites, we must stretch to accept and understand both sides fully. If and only if such Transcender projects can be done as complementary to the Sustainer focus of Life—environment value and quality for all people, then they pose no inherent threat.

Look again at Hubbert's diagram and the spike of very high quality energy lasting for only a brief moment in human history, and please accept for discussion purposes our framing of the necessity of the Transcenders from Chapter 1. Given these, we propose that from a species and planetary level, we should apply some of Hubbert's recommended "serious examination" to ask:

> How much of the unique and precious fossil fuel resource should be devoted to space exploration, travel, and colonizing, plus other Transcender interests, relative to efforts for cultural transition, regeneration, and sustaining Life on Earth?

One current thinker and writer on topics of sustainability and the environment, Nate Hagens, promotes the idea of the "Great Simplification" as a necessary coming transition and transformation required given our energy, mineral, and other resource and environmental limits (Hagens, 2018). This is an excellent notion, and it is compatible with the *Prosperous Way Down* (Odum and Odum, 2001) and the Great Transition scenarios. However, we think these types of futures—all under our umbrella heading of Sustainer projects and futures—are only about one-third of the story. Another one-third is the necessary Transcender project and future, and the final one-third is the relation between these two coupled complementary worldviews and action plans. We see it as critical that the relation between be mutualistic, complementary, synergistic, and cooperative rather than antagonistic.

If we consider that the Transcender mindset is normal, natural, valuable, and integral to Life—given that we have examples during every period in Life's history of growth, expansion, innovation, and transcending of environmental limitations and not only in the human species but also with all Life— then we could develop two modified scenarios. Here, we are addressing only the allocation of the fossil fuels in two additional scenarios to Hubbert's figure above:

IV. 50% of all the remaining fossil fuel energy supply is devoted to the Transcender programs of space exploration, space travel, and colonization of Life beyond Earth (and Earth-based Transcender projects). Or,

V. 100% of all the remaining fossil fuel energy supply is devoted to the Transcender programs of space exploration, space travel, and colonization of Life beyond Earth (and Earth-based Transcender projects).

Scenario IV makes sense if we consider an equal allotment of fossil fuels to both the Sustainer and Transcender projects. This type of material equality, as with philosophical aspects of equality, we see as necessary for cooperation to be possible.

Scenario V, while too extreme but used for discussion purposes, makes sense if we consider that since Sustainers (if successful) will eventually achieve a culture that runs on 100% renewable energy, this hypothetical group, worldview, or cultural type of people might as well begin this 100% renewable energy existence immediately.

If it seems odd that we of the self-proclaimed Sustainer worldview would advocate for the Transcenders to get either 50% or 100% of all the fossil fuels, remember what Peter Elbow suggested as the way he learned for breaking out of one system into a new and expanded one (a partial repeat of key Elbow quotes in Chapter 7):

> Searching for contradiction and affirming both sides can allow you to find both the limitations of the system in which you are working and a way to break out of it. . . .

> To actually get beyond that system you need to find the deepest contradictions and, instead of trying to reconcile them, heighten them by affirming both sides. And if you can nurture the contradictions cleverly enough, you can be led to a new system with a wider frame of reference, one that includes the two elements which were felt as contradictory in the old frame of reference. (p. 243)

We use this same method in what may seem a paradoxical capitulation or surrender to the Transcender camp—but note that this is "affirming both sides." We think it leads to a crucial insight, opening, and much greater potential that results from "embracing contraries" as Elbow urged, and seeking win—win as Covey recommended.

Our goal is to break out of the system we are (stuck) in so that we can resolve our systemic humans-in-the-environment problem. Rather than fighting adversaries, or seeking to reconcile the contradictions superficially, we, like Elbow, see value in first nurturing, heightening, deepening, and amplifying the contradictions.

Without "affirming both sides," understanding and seeing equal value in both sides—and this may be surprising to consider—we run the risk that we might get success on NEITHER front, we might fail at BOTH Sustainer and Transcender programs. The Sustainer program to sustain Life on Earth living within the real physical limits of Earth, and the Transcender program to transcend the real and physical limits of Earth, could both plausibly fail *if these two camps continue to fight and undermine each other*. Each a great Life urge; each also presents a formidable adversary and foe to any opponent. Neither will die nor surrender without a wicked battle. Thus, without consciously and intentionally affirming both sides, acknowledging the value and authentic authority of each view, we cannot hope for peaceful resolution and synergy of equals. We would get confusion, conflict, and what we have now—petty fighting, stalemate, no progress, severe, and potentially irreversible damage impacting both sides.

Miraculously, by accepting both sides, by letting go of either as a cherished or righteously superior worldview (Meadows' #1 source of leverage for change—the power to transcend paradigms) is perhaps the only way to open the potential that we succeed at BOTH Sustainer and Transcender programs, missions, narratives, and futures. If leading adherents of these camps can likewise find equality and acceptance, then we achieve the "super power" that comes with cooperation, coordination, and synergy. In a bit more specific terms, this would entail success with two seemingly contradictory programs:

1. Sustainer program success—healing environmental damage to Life support systems and transitioning to a complete reversal of human–environment relationship such that self-reinforcing and self-organizing processes enable the environment to improve in quality over time and Earth-based human culture to operate within the real physical and environmental limits of the Earth. We admit and note again that this is a huge, lofty, ambitious, and also plausibly attainable goal with prior Life precedent. And we add,
2. Transcender program success—transcending the resource and environmental limits of the Earth to allow continued expansion of human creativity, innovation, development, and advancement. This will involve beyond-Earth human culture that expands beyond the real physical, environmental, and spatial system boundaries of the Earth. It may also involve Earth-based Transcender success, perhaps generally related to transhumanism and genetic engineering, as examples. Yet another huge, lofty, ambitious, and also plausibly attainable goal (if we have faith in existing science and science fiction, human ingenuity, and the unquenchable drive to explore and expand; and if we get very skillful with Life–environment relations).

CASE STUDIES IN HOLISTIC LIFE TECHNOLOGY

Below, we briefly describe four leading examples (Table 8.1) of frameworks for technology and application that we see as compatible with holistic Life science. We also describe one generalized strategy for land development that illustrates

Table 8.1 Technologies Reviewed Here That are Compatible with Holistic Life Science

Cradle-to-cradle design
Biomimicry
Permaculture
Ecological engineering

how the foundations in this book lead to fundamentally different real-world applications. And, we describe in-depth how a focus on conversion of science enterprises and their facilities to sustainable operations is a crucial leading edge for the transformation wave front we propose.

As we examine these cases, we have in mind the ideal goal to avoid the systemic problem that Rosen identified with machine models and mechanistic technology—we seek to avoid the issue of adding one control loop, like a thermostat in a heating/cooling system, only to buy ever greater degrees of noise and thus an infinite series of control loops. The new kind of "craft" or "fabrication" as Rosen described, it will require a new combination of understanding and skill to design and construct systems that embody the kinds of self-making, self-healing, and self-improving capacities seen in organisms, ecosystems, and the biosphere. These new forms of knowledge and skill must draw on the primary Life value basis, all the principles and Life lessons in this book, and the awareness that science, technology, culture, and the environment are intimately entangled, co-creating, and interdependent in all places and at all times and timescales.

To add loops in a system is a common step. Given our holistic, long-term, and Life—environment quality goals, we need practical methods to gauge if adding a given loop leads to net gain and correct progress overall or leads to increased problems elsewhere which we are no longer allowing to be ignored or externalized. The loops we see as more effective than traditional control loops include autocatalytic, network flow, and feedback loops. Examples of the metrics we can use to gauge if a given system loop functions successfully include an assessment of net energy or energy return on energy invested (Hall, 2017). While this focuses on energy, it is holistic in accounting for all the energy that must be used, added, transformed, or invested to generate and supply the energy as input to the system. Estimates of net energy for various fuels and energy sources are very powerful for revealing just how high the quality of fossil fuel energy is, and to enable comparisons between various energy sources.

Another general way to gauge net Life benefit of a technology or application design, or a new system loop being considered, is the triple bottom line. This method accounts for costs and benefits to environmental, social, and economic assets for any given enterprise. We have already mentioned other holistic metrics that can be used to assess net Life benefit—Genuine Progress Index, ecological footprint, planetary boundaries, Index of Social Economic Welfare, Human Development Index, and Millennium Ecosystem Assessment. The Sustainable

Development Goals are another useful reference for describing crucial areas of performance for integrated social, environmental, and economic systems. And, we note again that impacts of an application should lead to building of functional gradients such as vertical gradients in soils (carbon, nitrogen, organic matter, etc.) and atmosphere (stratospheric ozone, greenhouse gases, etc.), maintenance of high biodiversity, and net gains in energy capacity.

Considering these many options for metrics and assessment tools, we see the need for collaborative effort to streamline and provide easy access to a comprehensive set of indicators. For example, the German Environmental Ministry publishes an annual report of key environmental indicators, which are used to direct policy initiatives (www.umweltbundesamt.de/en/data). In another approach, Kate Raworth has overlaid the planetary boundaries on a social foundation to create an indicator scheme referred to as doughnut economics (see Fig. 8.2). There are other creative indicators and tool kits that convey our current situation and trends. One could think of it as a set of readings on a "dashboard for Spaceship Earth."

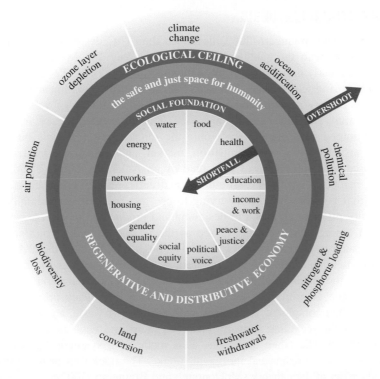

FIGURE 8.2

Doughnut Economics which combines upper bound for planetary limits and lower bound for social foundation.

Used with permission.

Or, for a bit less mechanical analogy, we might consider a set of vital signs for evaluating holistic health of Life—environment in its organismal, ecosystemic, and biospheric realms. Such vital signs would be a blend of those that are global and those that are customized for each unique local and regional context.

With the case studies we have chosen, we seek to point out crucial systems principles for holistic Life technology. For example, our first case study on cradle-to-cradle (CtC) design is based on a strategy of design and manufacturing in which the construction and deconstruction processes are designed, developed, and implemented together, as interdependent and mutually supporting in a single unified production-use-reclamation cycle. Thus, this specific example system is directly compatible with and borrows from the generic Life—environment principle of coupled transformers and coupled complementary processes in which the "composer" and "decomposer" functions co-arise, cooperate, and coevolve. We next examine more details on this case and the others.

CRADLE-TO-CRADLE DESIGN

McDonough and Braungart (2002) described their CtC design process in a book by that name, with the subtitle "Remaking the Way We Make Things." This approach is clearly in response to the standard paradigm of cradle-to-grave, which in manufacturing terms means raw materials to landfill. This linear thinking has led to unnecessary resource acquisition and the accumulation of waste piles that otherwise may still have useful capacities. This is another legacy from a time when resources were plentiful (not only the mineral raw materials but also the energies to extract, transport, transform, market, and trash those materials). The CtC, closed-loop approach was promoted by an American architect and German chemist, respectively. These two authors and entrepreneurs proposed a coherent strategy that goes beyond merely trying to be "less bad" environmentally toward "a radically different approach to designing and producing the objects we use and enjoy, an emergent movement we see as the next industrial revolution" (p. 6). Their visionary approach starts with McDonough's realization that "design is a signal of intention." The two collaborate and add to this awareness the intention to "love all the children, of all species, for all time." This clearly is aligned with our Life value as primary basis for thought and action. They propose a design philosophy that completely does away with the idea of waste, and they base this potential on the recycling capacity observed in natural ecological systems. The book itself embodies their approach—it is made of a synthetic "paper" that is "made from plastic resins and inorganic fillers." The book thus proves the practical application of their theory (McDonough and Braungart, 2002):

> This material is not only waterproof, extremely durable, and (in may localities) recyclable by conventional means; it is also a prototype for the book as a "technical nutrient," that is, as a product that can be broken down and

circulated infinitely in industrial cycles—made and remade as "paper" or other products. (p. 5)

Their proposed and well-developed system includes conceptual distinctions between biological nutrients and technical nutrients, and the way these flows are processed in ecological and industrial cycles, respectively. This clarification and priority issue corroborates Rosen's distinction between Life and machine and our recommendation to be fully aware of the models and metaphors we use and their ultimate impacts.

Operating via McDonough Braungart Design Chemistry at mbdc.com, these two continue to collaborate and have built a large organization that includes CtC certification process (with variable ratings attainable, Silver, Gold, etc.) Their stated philosophy reads (MBDC, 2018):

> Cradle to Cradle encourages us to step back from the routines of daily problem-solving and rethink the frame conditions that shape our designs. Rather than seeking to minimize the harm we inflict, Cradle to Cradle reframes design as a beneficial, regenerative force—one that seeks to create ecological footprints to delight in, not lament. It expands the definition of design quality to include positive effects on economic, ecological and social health. Cradle to Cradle rejects the idea that growth is detrimental to environmental health; after all, in nature growth is good. Instead, it promotes the idea that good design supports a rich human experience with all that entails—fun, beauty, enjoyment, inspiration and poetry—and still encourages environmental health and abundance.

As with many case studies, and the complex ideas in this book, language and ideas are complex and require highly developed understanding and skills for integration, synthesis, and evaluation. We can accept McDonough and Braungart's promotion of "growth" above given that the growth is of the same holistic form as growth in "nature" as in Life. In this context, growth is a regenerative process, replacing and renewing a previous system, not an increase in accumulation. Furthermore, it is always accompanied by use of renewable energy, recycling materials processes, and net improvement of Life and its environmental context. If this is true, then we can agree that kind of growth is good.

McDonough and Braungart and their CtC design have partnered with very large corporations such as those in textiles, skin care, food and beverage, packing, and more. As such, their impact and reach is great. Whether their incremental steps to improve industrial design and product cycles can scale up to systemic change for environmental sustainability remains to be seen. It may be aided in success by some of our next case studies.

BIOMIMICRY

A term popularized and promoted by Benyus (1997), biomimicry, is what its name implies—an effort to mimic biology. The Biomimicry Institute website provides additional information on the intention toward "innovation that seeks sustainable solutions to human challenges" and their strategy of innovation based on mimicry:

> A sustainable world already exists.
>
> Humans are clever, but without intending to, we have created massive sustainability problems for future generations. Fortunately, solutions to these global challenges are all around us.
>
> Biomimicry is an approach to innovation that seeks sustainable solutions to human challenges by emulating nature's time-tested patterns and strategies. The goal is to create products, processes, and policies—new ways of living—that are well-adapted to life on earth over the long haul.
>
> The core idea is that nature has already solved many of the problems we are grappling with. Animals, plants, and microbes are the consummate engineers. After billions of years of research and development…what surrounds us is the secret to survival.
>
> **Biomimicry.org (2018)**

The Biomimicry Institute provides many educational resources on biomimicry, they sponsor and support the Biomimicry Educators Network, and they have a fellows program for faculty and administrators. A few of the specific examples they promote as proof of the concept, from their website:

1. Learning from humpback whales how to create efficient wind power. At 40—50 ft long and 80,000 pounds, they say "the whale's surprising dexterity is due mainly to its flippers, which have large, irregular looking bumps called tubercles across their leading edges." They explain how this incidence of mimicry has spawned a commercial enterprise, WhalePower, a company that "is applying the lessons learned from humpback whales to the design of wind turbines to increase their efficiency…."
2. Learning from prairies how to grow food in resilient ways. A prairie they assert, like "any natural ecosystem [is] a remarkable system of food production: productive, resilient, self-enriching, and ultimately sustainable." The example they provide overlaps with our links from permaculture below. The Land Institute, according to Biomimicry.org:

> …has been working successfully to revolutionize the conceptual foundations of modern agriculture by using natural prairies as a model: they have been

demonstrating that using deep-rooted plants which survive year-to-year (perennials) in agricultural systems which mimic stable natural ecosystems — rather than the weedy crops common to many modern agricultural systems — can produce equivalent yields of grain and maintain and even improve the water and soil resources upon which all future agriculture depends.

3. Learning from termites how to create sustainable buildings. Closed, cramped, and often found in very hot climates, termites have developed sustainable methods for cooling their homes. The website reports one case where this has been mimicked successfully:

> ...the Eastgate Building, an office complex in Harare, Zimbabwe, has an internal climate control system originally inspired by the structure of termite mounds. Further research is revealing more about the relationship between mound structure and internal temperature and could influence additional building designs as our understanding grows.

The Biomimicry Institute conducts an annual design competition, the Biomimicry Global Design Challenge, and an annual entrepreneur competition, the Biomimicry Launchpad, which they say "is the world's only accelerator program that supports early-stage entrepreneurs working to bring nature-inspired solutions to market." They also state (Biomimicry.org, 2018):

> We need proven, focused, regional solutions to address global social and environmental issues, like those in the United Nations' Sustainable Development Goals. Biomimicry can unlock new insights for how to develop a more resilient future for people and planet.

At the end of the Launchpad competition, one team is awarded $100,000 to support their project, funding provided by the Ray C. Anderson Foundation.

There are numerous exciting and unexplored ways that biomimicry can be further integrated into design and engineering of human systems. Our only criticism of this approach is the emphasis primarily on organismal innovations from nature and less attention to mimicking processes at ecosystemic and biospheric Life—environment systems scales. However, the Biomimicry Institute already cites the example above of mimicking prairies to redesign agriculture based on natural ecosystems. To integrate the ecosystem and biosphere scales more fully, as with our three holons and the hyperset formalism, would be toward an allied approach we might call "ecomimicry." This new line of research is open and waiting for readers to further develop it.

PERMACULTURE

There is a noticeable sea change in Americans' awareness of alternative agriculture and food, characterized by the rise of organic farming, local food movements such as community supported agriculture, the importation of the slow food and real food movements, and the popularization of food systems by authors such as Michael Pollan and Dan Barber, to name a few. This awareness encompasses human health, animal rights, and environmental concerns in one fundamental interaction that humans have with nature. In fact, it is through consumption, metabolism, and egestion, that the inseparable embeddedness of nature in humans becomes evident, which we try to capture elsewhere in terms of a hyperset formulation. Therefore, growing and eating food is one area that has profound ability to capture holistic Life—environment principles. One major historical transition in agriculture is when humans moved away from polyculture based on perennial plants to monoculture based on annual plants. The annual plants reward the farmer with greater yields (investing energies in larger seeds rather than longer roots) but are more needy in terms of nutrients and do less to build and maintain soils.

The modern farming approach called *Permaculture* is returning to a way that promotes polyculture and perennials, with the added benefits of advanced techniques to retain high yields. The term comes from a combination or "permanent" and "agriculture." Bill Mollison started this holistic practice with the idea to produce food like natural ecosystems he observed studying rainforests in Tasmania. Mollison (1996) wrote of the strongly ethical motivation for his proposed form of sustainable human—environment relation:

> The Prime Directive of Permaculture
> The only ethical decision is to take responsibility for our own existence and that of our children. **Make it now.**

This concise principle fits with the other holistic Life ethics we have seen including Schweitzer, Leopold, and more.

The community of permaculture practitioners has also spread to include a global network of training and certification programs. Many have successfully implemented permaculture and developed off-shoots and allied extensions inspired by permaculture, including Alan Savory's Holistic Management, Joel Salatin's Polyface Farm, and the Land Institute in Kansas (which was cited above as a biomimicry example). Permaculture applies to other land uses as well as food production and improves the sustainability of homes, communities, landscaping around buildings, and more. In each case, those employing the permaculture principles balance specific efforts tailored to the details of the local natural community-ecosystem with universal principles of sustainable and win—win Life—environment relations.

ECOLOGICAL ENGINEERING

In some ways, similar to the biomimicry approach, a group of progressive, ecologically-oriented engineers have set out to include ecological principles into engineering projects. The emerging field of ecological engineering is a fusion field that holds great promise for human applications given its basis in ecological science. For example, rather than building a physical infrastructure wastewater treatment plant, a constructed wetland can serve the same purpose at lower cost and more in tune with natural biogeochemical processes. We discuss only the approach of Pat Kangas at the University of Maryland for our specific case study. Other leading faculty and centers include Bill Mitsch at Florida Gulf Coast University, and Bhavik Bakshi works in the closely allied field of industrial ecology.

In the preface to his 2004 textbook for graduate level study, Pat Kangas gave this description of ecological engineering (Kangas, 2004):

> The Earth's biosphere contains a tremendous variety of existing ecosystems, and ecosystems that never existed before are being created by mixing species and geochemical processes together in new ways. Many different applications are utilizing these old and new ecosystems but with little utility, yet. Ecological engineering is emerging as the discipline that offers unification with principles for understanding and for designing all ecosystem-scale applications.

Kangas credits H.T. Odum's teaching for his writing of this book of principles and practices of ecological engineering. In his textbook, he covers three major principles: the energy signature, self-organization, and preadaptation. He describes these principles:

> Energy signature: The set of energy sources or forcing functions which determine ecosystem structure and function.

> Self-organization: The selection process through which ecosystems emerge in response to environmental conditions by filtering of genetic inputs.

> Preadaptation: The phenomenon, which occurs entirely fortuitously, whereby adaptations that arise through natural selection for one set of environmental conditions just happen to be adaptive for a new set of environmental conditions that the organism had not previously been exposed to. (p. 17)

Using these three major principles, Kangas examines ecological engineering as applied to wetlands designed for water treatment, soil bioengineering in urban and agricultural systems, restoration ecology of saltmarshes and artificial reefs, and control of and learning from exotic species. Kangas' faculty web pages at the University of Maryland describe many other ecological engineering projects he has done, including a floating solar-powered lake restoring system to improve water quality (Yaron et al., 2000), and the Greenhab Project of the Mars Desert Research Station (MDRS) including "the construction and implementation of a new greenhouse, an experimental living machine for wastewater treatment and a recycling loop that returns treated greywater back to the MDRS habitat's toilet" (Blersch and Kangas, 2003).

His major principles and his case studies in ecological engineering share many of the principles and concepts we have promoted, including anticipatory approach, the importance of how we draw a system boundary, the ecosystem as crucial model system, and more. The MDRS project helps to demonstrate how the holistic Life science and technology we see as effective for sustainability on Earth is likewise helpful for Transcender and space efforts. While Kangas, John Todd, and others use the term "living machine" for their ecologically engineered systems, we recommend keeping these as categorically separate, again following Rosen and also McDonough and Braungart. We admit that machines can have life-like and Sustainer qualities, such as running on renewable energy and recycling materials processes. And, machines can be made to mimic other Life capacities. Humans can also design and embed machines and the associated construction, use and recycling processes within a Life-valuing culture as in industrial ecology.

Many other inspiring and innovative examples exist, but we cannot cover all types of the great diversity now arising. One last hopeful example combines environmental education for children and teachers with the highest standard for sustainable buildings. The Alice Ferguson Foundation recently constructed a meeting and educational building that conforms to the Living Building Challenge criteria. This augments their transformative hands-on experiences, experiential education, and agriculture education with a feature of their built environment that demonstrates for the children, teachers in training, and visitors how to "walk the walk" and lead by example. Their website states:

> The net zero energy, net zero water, carbon neutral, and non-toxic component requirements of the Living Building Challenge™ will not only enhance and upgrade our structures but also serve as tools for teaching Science, Technology, Engineering, and Math (STEM) concepts, as well as augment our core ecological curriculum.

And

> Living Buildings are designed to function like species in an ecosystem and mimic the beauty, resourcefulness, and efficiency of nature's architecture. Living Buildings are designed to regenerate, not deplete, their surroundings.
>
> **www.fergusonfoundation.org**

The Living Building Challenge comprises an even more stringent set of design and functional sustainable building criteria compared to LEED standards. Case studies of certified Living Buildings can be found online at https://living-future.org/lbc/.

A HYPOTHETICAL APPLICATION—HOLISTIC LAND DEVELOPMENT

Our next case study is a generalized, hypothetical, and potential application. The example also shows a case of reversal of approach relative to the existing conventional method that fits with the Great Change, paradigm shift, and sea change we have suggested is necessary. This is a contrast of two approaches to land development that we presented in a prior article (Fiscus et al., 2012). In a nutshell, the current practice and recommended alternative approach to land development are as follows:

1. Current practice—add people and built environment to the landscape and simultaneously remove, degrade, destroy, or diminish natural systems that are Life support at that site
2. Alternative holistic Life practice—add people and built environment to the landscape and simultaneously add, enhance, increase or regenerate natural systems that are Life support at that site

We described the crux of the distinction before (Fiscus et al., 2012) as linked to our founding principles of drawing system boundaries wisely, and the unfractionable Life—environment relation. We wrote that the critical issue is

> ...how we split human life from environmental life support in thought and action is that when we look at a parcel of land, we are able to think of it as occupied by either humans or some natural system.

A bit later we described the alternative mental model and how it radically transforms how we develop land:

> Now imagine that we instead operate from the proposed unified paradigm. In the revised scenario, we understand and treat life and environment as necessarily unified and interdependent. Thus, our "unit of development" shifts in systemic fashion—instead of a focus merely on units of land, or buildings, corporations or people, we must also integrate unit-models at the ecosystem and biosphere scales to actualize continual attention and value for energy flow, materials cycling, biodiversity, primary production, decomposition, soil formation, atmospheric regulation and other essential ecosystem services which all people need for life.

We then explained how the holistic Life science and holistic Life applications views integrate Life and environment:

> ...when we look at a parcel of land, we are NOT able to think of this unit of the environment as occupied by either humans or some natural system but are only able to think of this unit as necessarily occupied by both humans and some natural environmental system.

This modified approach, compatible with employing the hyperset formalism of Chapter 6, prohibits fragmentation of Life—environment, and in that holistic mandate also prevents any externalized damage to environmental Life support. This anticipatory approach in turn prevents a debt that someone else, perhaps in some other place in the world, or at some future date in time, must pay for the damage done to a Life—environment system ultimately more valuable (since Life value is fundamental and is of the greatest value) than any short-term profit or even functional benefit (housing, business, etc.) gained on that landscape location.

Scale becomes important in this new approach. The land, whether we occupy it or not, has a certain productive capacity depending on the climate, rainfall, geology, and biodiversity, etc. When we see ourselves as integrated into a specific place, our aim becomes to maintain and promote that productivity for ourselves. In this case, our interest aligns and also supports productivity for the larger Life—environment system. A main pathology that interferes with this approach is the loss of connection to place, the ease of portable capital, and a growth-oriented paradigm that promotes short-term exploitation over long-term stewardship. The importance of scale is not simply the spatial one from local to global, but also a temporal one from short term to long term. Societies or groups that easily forget the past are likely to also ignore the future.

Looking at the Chesapeake Bay watershed and its regional landscape, as just one example of many in the United States, we see the ultimate impact of the current development paradigm based on the current mainstream paradigm of science and life science. In terms of environmental quality, carrying capacity, and sustainability, the Bay and its watershed suffer from chronic and systemic environmental degradation (IAN, 2012; Hyslop, 2012; Town Creek Foundation, 2018), and the costs for restoration and remediation are so high that governments and environmental groups have stopped publishing estimates.

To achieve a different ultimate outcome for land and environmental quality, as development plays out, if we plan to add people, we now know we must add more Life support capacity, not reduce it. Instead of being removed, the necessary environmental complement to people must be grown, developed, and nurtured, intentionally. And, the integrated development of land for people and Life-support must be done in concert and in close coordination. This necessary integration—a form of internalization that is the exact opposite of externalization—may be accomplished by adding Life support capacity at local, regional, and/or global scales. However, holistic Life science and technology makes the accounting and the actual implementation essential. Policies and regulations which require no net loss of wetlands serve as a compatible example, but we assert this approach must be extended to ensure no net loss and even increase and enhancement to match population growth, of the full suite of essential ecosystem services and Life-support capacity.

FOCUS AND PRIORITY CASE STUDY—SCIENCE FACILITIES

A central theme and reason for being of this book is science reform. Just as we have attempted to present coherent value, conceptual, and methodological foundations for holistic Life science, we assert that sustainable physical foundations for scientific operations are equally essential. Science supports most if not all of the other example technologies and applications for sustainability we have cited—design, manufacturing, agriculture, engineering, buildings, land development—and in order to provide consistent, effective, and long-standing support for all these and sustain itself, science facilities in academia, government, and nonprofit sectors must transform themselves first in an effort to lead by example.

In the Conclusion section of his book, *Walden*, Thoreau (1854), one of the early American environmental thought leaders, wrote:

> If you have built castles in the air, your work need not be lost; that is where they should be. Now put the foundations under them. (p. 215)

To reimagine and then transform science, Life science, environmental science, and related facilities to run fully on renewable energy, integrate operations within recycling materials processes, adapt, and coevolve with the local environment, yield net improvement in air, water, soil, biodiversity, and atmosphere quality and the other win−win relational impacts we have described, admittedly, entails great challenges. Science enterprises require large amounts of energy, often employ toxic substances, and use high-tech materials like computers and other equipment that can be very hard to recycle. However, if science is serious about assisting and even leading the way forward to lasting and systemic solutions to the human−environment crisis, then we have no alternative. To follow the six foundational Life science principles—to be anticipatory and holistic, to affirm Life value, to internalize our own impacts, to embrace complexity fully, and to be radically empirical—is to get to the root of the matter and help industrial culture reverse our current unsustainable trajectory.

Thoreau also wrote in his *Journal* (Walls, 1999):

> There is no such thing as pure *objective* observation. Your observation, to be interesting, i.e., to be significant, must be *subjective*. The sum of what the writer of whatever class has to report is simply some human experience, whether he be poet or philosopher or man of science. The man of most science is the man most alive, whose life is the greatest event. (p. 60−61)

And

> As science which is poetry *professed* by the civilized state—measuring the unfathomed with its telescope—& microscope—but feebly and partially—we want something more comprehensive & assertive which may be called conscience perhaps—and signify a practical growth. (p. 9)

Here, we find insights from a radical and pragmatic thinker who lived and worked before the word "scientist" even existed. Thoreau adds poetic energy to our previous calls for the integration of ethics with science, for the unity of subjective and objective perspectives, and for a science that aspires to and lives up to the highest of standards. Surely, such ideals call for realization in our facilities functional and elegant sustainability that matches the greatest of our insights, knowledge, and wisdom gained from the centuries of science in practice.

Prior chapters have described how we need to change the "idea-tools" we think with, to transform our science paradigm, to integrate the most complex advances in theoretical modeling, and simultaneously to move toward Leopold's "thinking like a mountain" or Lappé's Ecomind. Science is very much like the collective mind of society, and as such it must be clearly focused on what matters most in order to best serve people and Life itself. But, we also know from Rosen that formal systems of science process and active fabrication or realization go hand-in-hand and co-create each other. Construction occurs in the mind (ideas) and in real world (publications, experiments, measurements, etc.), simultaneously. Contrary to the common lament, "We can't change the system," we see evidence everywhere that we can't help but to change the system. With every thought and action, or thought—action, we change ourselves and the world in a seamless process.

Donella Meadows was aware of the imperative for science to lead by example, and she wrote of the need for what she called a "think-do tank" to replace the idea of a think tank as focused on abstract ideas or science disconnected from grounding in reality. Thus, she and partners created Cobb Hill, a co-housing community and organic farm integrated with the Sustainability Institute so that the thinking and the doing would be in very close physical and cooperative communication. Some of the organizations she founded have changed names since her death, but her holistic systems and modeling work, leadership, and legacy of publications continue to teach and to serve us in our efforts.

An ambitious and creative cohort of other institutions provides a contagious sense of success that this transformation to sustainable science is not only possible but yields multiple great rewards. Here, are a few of which we are aware.

Schumacher College in England, led by Stephan Harding and with visiting scholars like Fritjof Capra and more, offers one of the few graduate degrees in Holistic Science. Their facilities at Dartington likewise model sustainable practices and help the scientists, teachers, and students participating to experience mutually beneficial relationships to the land for growing food and to nature for renewing the spirit.

The College of the Atlantic in Maine offers just one bachelor's degree—Human Ecology. One of the colleges and universities in the United States leading the change to sustainable operations, they have been climate neutral for several years. They achieved this by reducing energy usage, utilizing renewable energy, and offsetting emissions through verified offset programs.

At Pennsylvania State University, in the 1990s, professor Christopher Uhl was instrumental in studying the environmental performance of his own university. On his faculty website, he reports that he "used 'sustainability indicators' to track the performance of Penn State in areas such as water use, energy consumption, waste generation, and recycling efficiency." This work led to his publication in the journal, *Conservation Biology* (Uhl et al., 1996), in which he and his colleagues described sustainability as "a touchstone concept for university operations, education, and research." Today, Penn State hosts the Sustainability Institute, which since its founding in 2013 has grown to support an extensive staff, associated faculty, student-led initiatives and multiple efforts in education, research, and campus operations. Uhl has continued his work in this area and has published books on developing ecological consciousness (Uhl, 2003) and reform of education culture to support "teaching as if life matters" (Uhl and Stuchul, 2011).

At the University of New Hampshire (UNH), sustainability is promoted as a key strength on the main web page and at the set of pages devoted to sustainability. Here, one quickly learns of UNH's leadership role and clear commitment via a set of statistics; for example, UNH:

1. is one of only three US institutions of higher education to earn an AASHE STARS Platinum rating,
2. has a main campus that is 100% powered by renewable energy,
3. is home to the first endowed sustainability institute at a university in the United States, and
4. has the first organic dairy research farm.

In 2017, UNH introduced the Sustainability Indicator Management and Analysis Platform (SIMAP). SIMAP is a resource available to any college or university, and "is a comprehensive footprint reporting tool designed for campuses" that provides campuses "a simple and affordable online platform for tracking, reporting, and managing their carbon and nitrogen footprints."

Aber et al. (2009) published a book about the community and the process of UNH's efforts to become more sustainable.

The stories of self-transformation at colleges and universities, and the innovative and concerned leaders who have charted the course of change, has many more chapters we cannot cover here. David Orr at Oberlin, a team at Chatham College—alma mater of Rachel Carson—and many more are worth finding and learning from their experiences.

Inspired by these champions of sustainable leading by example, we believe that scientists are integrated with and serve as representatives for their institutions and facilities. Scientists and their science enterprises cannot continue to provide consistent and trusted leadership if we act as if we are special or have some exemption from the imperative to be sustainable and value Life. Such a claim of exceptionalism can be found tacitly if not explicitly with many mainstream sectors of society, organizations, and individuals. In essence, they are saying "I am allowed to be unsustainable because my work in my sector is so important." This argument, if allowed anywhere, must be allowed everywhere. Thus, it cannot be allowed—it is a false dichotomy and results in dangerous demotion of Life value relative to lesser values. Each sector, field, discipline, and enterprise must accept the new Sustainer terms—do whatever you do AND improve the environment as you do it. Imagine the growing sense of pride and accomplishment as we are more and more able to say and show—"We do science in ways that value Life. And, here is how we do it."

CHALLENGES TO HOLISTIC TECHNOLOGY TRANSFORMATION

Science facilities, food production, buildings, manufacturing, and other forms of technology and culture can and must be transformed to improve rather than damage environmental quality over time. We have shown how this can be done effectively, and many case studies and success stories support this view. We also acknowledge challenges, complexity, and inertia that must be overcome.

One challenge has to do with technologies that impact human life and survival. The heart valve that we mentioned in Chapter 7, for example, raises interesting dilemmas. We described this technology as beneficial and a case where applying mechanistic models to living beings makes sense. This acceptance and positive view was based on (1) the similarity between a human heart and a mechanical pump and (2) the clear and profound impact on extension of lifespan and improvement of quality of life. This assessment was done with a relatively narrow system boundary—we only considered the life of one person. If we expand the system boundary and use a more holistic approach as we have said is necessary, then the story becomes more complex. Is this technology still a net positive application when we consider all people and their social and environmental context? If my father gets a heart valve, but the energy, money, expertise, and other costs of his heart valve prohibit, impede, or delay three other people from getting the same surgery, is it truly sustainable? Does it lead to a real net increase and benefit in Life value? In addition to considering the ecosystem services involved, we here also integrate economics and social justice as in Agyeman's "just sustainabilities." Social justice is equally as important to consider as the Life—environment sustainability concepts. Thus, while we do feel strongly that sustainable technology is necessary and possible, we do not suggest that it will be easy to implement at the level of all industrial culture.

In global industrial culture, we are constantly confronted with new books, scientific publications, videos, websites, and other media where innovations and ideas are promoted. To digest this massive flow of information can feel like trying to sip from a fire hose. Add to this the pace of life and the constant rush and it becomes difficult to make time to reflect and consider fully which ideas, innovations, and technologies can be traced back to solid foundations such as those based on Life value and holistic Life science. We hope that the primary distinction between Sustainer and Transcender perspectives, and the other criteria in this book can make this process more practical and streamlined.

Several counter examples, some of which represent negative technology to us, are worth mention. We do not propose to evaluate good versus bad technology from the Transcender perspective. We do see it as necessary to always assess clearly whether a new technology or conceptual approach abides by a primary Life value basis. For example:

1. Ray Kurzweil's "singularity" ideas. The techno-optimist vision of exponential growth in computing nanotechnology and robotics, that foresees a merger with humans so that humans "transcend biology", is clearly part of the Transcender program as we have depicted it. This, and other Transcender ideas and technology, could be beneficial or at least neutral if and only if it does not conflict with or hinder the Sustainer program. We see the next and more important Life—environment dynamical system singularity as a bifurcation into Sustainers and Transcenders.

2. Walmart drones for pollination. This appears extremely infeasible, misses the core Life principles and lessons, and could do great harm. Such a mechanical model applied to such an elegant biological and ecological context would entail not just adding one extraneous control loop but millions. This evokes the concerns with infinite regress that Rosen warned about, and we anticipate increased energy use and Life—environment damage. Overall, this is not a wise alternative compared to protecting the biodiversity and healthy environments of our natural pollinators. See Hetherington (2018) for news coverage of this idea.

3. Nanotechnology developed and involving manufacture of products and materials without a coupled complementary decomposition process. Unless a "CtC," full circle, composer—decomposer system as in Life is employed, this runs great risk of poisoning essential Life-support systems and is irresponsible. To use the precautionary principle, and the principles of anticipatory science and primary Life value necessitates the decomposition process be integrated so that no nano-materials escape or accumulate.

4. Geo-engineering projects mentioned earlier. Seeding oceans with iron or limiting nutrients to increase algal growth and sequester CO_2, for example. Given the scale on which many of these have been proposed, and the complexity of ocean ecosystems, a valid assessment of the multiplicity of intended and unintended consequences seems impossible. If we cannot have confidence that a technology will lead to net benefit for Life, then we should not develop or employ it.

While sustainable and regenerative net impacts of technology are possible, these and many other challenges exist for achieving full transformation successfully. The money system, which as now designed is dependent on growth and thus on negative forms of economic growth (resource extraction, etc.), is another crucial obstacle. Goerner et al. (2009), Hornborg (2001, 2011), Lietaer et al. (2012), Mellor (2010), and others have put forward practical ways to reform the monetary system that would support the transition to holistic Life technology. The complex arena of human values is another area of challenge and change needed—through better environmental, ecological, and holistic Life education and other means, we must build consensus that Life value is the primary basis for value. The need for maturity and complex self-awareness, beyond simple either/or, true/false thinking is another challenge. The holistic Life science and story we have developed in this book promises a better future for ourselves and our environment, but only if the human sense of self evolves to embrace both a discrete self and a sustained self. This entails a new acceptance of death as a necessary aspect of the larger, integrated Life—environment system by which individuals, organisms, and other discrete Life forms, die, decompose, are recycled, and feed other Life forms and soils so that Life itself, Life as a unified whole, can learn and improve from generation to generation. We predict that as holistic Life applications and technologies increase the pace of change should accelerate. The multiple benefits to human quality of life; environmental Life support; and sustainable and resilient food, water, and other ecosystem services reduced costs, and thus, more resources available will provide feedback and incentive to further increase research, development, and new applications.

Sustainability: A goal for all

INTRODUCTION

We started this book with the goal to describe science reform efforts to help solve the sustainability crisis. Remember—some change will be required, as, by definition, an unsustainable system cannot continue along its current ways. Whether the change is planned and orderly or forced and chaotic depends on the level of anticipation and effort to be proactive. Our working assumption and intuitive insight from a combined 50 years of research and applications in systems and network ecology was that the existing science paradigm and analytical methods based on it have not worked. Thus, from that starting point, in order to address the crisis, we did the opposite of analysis—we enlarged the problem context rather than picking out a reduced subset of the system-of-systems complexity to tackle in isolation.

The previous chapters have sought to present an alternative holistic, Life-centered, non-mechanistic science as a complement to existing analytical, reductionist, and mechanistic mainstream science. We have sought to build the foundational framework for holistic Life science with a coherent, logical, stable, and long-lasting organizational structure. The layers for this structure began with values. We showed that a conscious choice to set Life as the primary basis for value has the power to form the most universally relevant and thus the most stable foundation for Life science and the methods, technology, applications, and policies that can later flow from the Life value foundation. We then described six founding principles of holistic Life science and seven Life lessons and methods developed from past work, which are relevant for future work. We also developed the coherent narrative, again corroborated by workers in diverse disciplines, that the systemic environmental degradation we now experience has been caused largely by the dominant mechanistic root metaphor at the heart of science. Lastly, we described several existing allied works and innovative leaders who are already implementing technology and applications that are compatible with the science we propose as well as a hypothetical holistic approach to land development based on our framework.

In this chapter, we address several additional conceptual and scientific perspectives to help further illustrate the framework and better weave together the many complex threads, ideas, and methods. These final thoughts seek to build yet

Foundations for Sustainability. DOI: https://doi.org/10.1016/B978-0-12-811460-5.00009-1

another bridge linking the holistic Life science paradigm to the everyday realm of common sense, conventional wisdom, day-to-day activities, and intuitive sensibilities by which we normally think about Life, environment, human culture, and the myriad relations between. As a final set of parting words, we hope this chapter serves to inspire hope and optimism that the reformed scientific foundations we have presented hold true potential to catalyze lasting and systemic change for human—environment sustainability.

TWO SUCCESSFUL FUTURES ARE BETTER THAN ONE

Our holistic and systemic approach began by looking at what may be the most difficult and seemingly intractable conflict in our current crisis. Instead of continuing the conflict by trying to argue for superiority or primacy of the Sustainer worldview and program, we sought to "embrace contraries" and "amplify both sides" in the spirit of Elbow (1986) in order to break out of the stuck system we are in and to help bring into being a new system in which these two seemingly contradictory worldviews—sustainers and transcenders—can be mutually beneficial and complementary. This approach also follows the thought leadership of Stephen Covey and his principles to "start with the end in mind" and "think win—win." By employing the concepts and methods in this book, we see it as practically achievable that we can indeed succeed on both future paths that both Sustainers and Transcenders can win. We have tried to portray how these two groups can, and in fact must, help each other achieve Great Change, a Great Transition, and ultimately Great Life success. A stark way to frame our current moment in history is that we can choose to succeed together or fail separately—united we stand, divided we fall.

In our 2015 book, *Flourishing Within Limits to Growth*, we presented a case and a collaborative vision by a team of systems and network ecologists showing how by "following nature's way" we humans can exist with rich lives and culture while also preserving environmental quality. That book has many recommendations and practical steps that are compatible with this book. These include renewable energy, recycling materials, Pigovian taxes, investments in education and family planning, and the promotion of research, development, and innovation. We continue to believe those recommendations taken together are essential for a "system of solutions" solving the entangled "system of problems" that comprise the human—environment crisis.

The main ideas we have added and expanded upon in this book include an explicit foundation of the Life value proposition and to embrace the opposite worldview, belief system, or system of ideas relative to the human—environment interface. *Flourishing* primarily addressed the perspective of the Sustainers, while in this book, we acknowledge the equal validity of their complementary counterparts, the Transcenders. Our ambitious ideal goal is assistance so that both camps, thus ideally all humans in a leadership and stewardship role aiding all Life, can flourish—the Sustainers flourishing on Earth within the physical, material, spatial, and Life support (carrying capacity) limits to growth, and the Transcenders flourishing on the different yet necessarily cooperative project to grow and develop beyond these environmental and planetary limits, whether on Earth (for certain types of non-material growth) or beyond (for colonizing Life in space and on other worlds).

Our choice to reverse the trend and habit of continual splitting, reducing, fragmenting, and analyzing is intentional. We see the imbalance and extreme bias toward the analytical mode as very much part and parcel with the root cause of environmental degradation that is the glaring signature and unintended negative consequence of industrial culture. Going back to the origin, we have shown it possible to reverse the standard biology textbook approach that separates life from environment when describing the original and fundamental nature of Life. And, in the present moment of our global community, we choose to unify rather than further polarize differences between people based on cherished ideas, worldviews, politics, and ideological "tribes." This is not an unidentified third way option that looks to reconcile all dissimilarities, but rather a way forward that retains the distinctions and advantages of both approaches. It is not a new discovery that a functioning dialectic—where two sides continually oppose each other—can be stable particularly when there is balance in ascendency of the two sides. The environmental crisis is one manifestation of the asymmetric imbalance toward Transcenders and away from Sustainers, and toward reductionism away from holism. Our book aims to reposition and stimulate the role of Sustainers and holism, but not to eliminate the opposites.

The ultimate destination we foresee from an endless series of splitting, fragmenting, and division is that we will cut ourselves off from each other and from the Life support context of the world, and then we will cease to exist. The evidence for a trend of such impacts and consequences is already here and abundantly clear, and every crucial aspect of our shared planetary home most valuable for sustaining Life is under threat of destruction and has shown quantifiable damage for decades. Add to these ills the increasing economic volatility and systemic risk, increasing military and civil conflict within and between nations, and it is overwhelmingly clear that we need radical solutions with power to transform mindsets, to heal land and people, and to bring people together.

Thus, we must change course. We have shown the need for a compensating focus on unifying and bringing together, which, after a corrective transition period, can lead to a balanced scientific program of equal parts analysis and synthesis to help foster a balanced culture. In this new era, we imagine we will achieve an organic unity and empowerment by which we and our planetary environment co-develop and co-evolve in mutually beneficial synergy such that human Life and Earth environment both improve in quality over time. Does this sound naively optimistic and idealistic? Perhaps, from the old mechanistic mindset and current framing. But we know now it is the trajectory that Life on Earth has followed during its long-term development as evidenced by increasing its complexity, diversity, and interdependencies. Why should adding humans be so different?

The path of unifying and wholeness, as we have described several times, can be supported with evidence, results, and methods of science, and it can be corroborated by perennial philosophy, wisdom traditions, and Eastern philosophies and religions. Thich Nhat Hanh, a Buddhist monk from Vietnam, has written about "inter-being" as one of the most important teachings of Buddhism. He has written "We are here to awaken from the illusion of separateness." The radical empiricism of Thich Nhat Hanh, other Buddhists, Native Americans, and people with diverse viewpoints can serve to disrupt our old habits and ways of thinking just as we have sought to do with new holistic science perspectives. Take, for example, this approach to understand the fundamental nature of things such as a flower:

> There is no permanent entity within us, there is only a stream of being. There is always a lot of input and output. The input and the output happen in every second, and we should learn how to look at life as streams of being, and not as separate entities. This is a very profound teaching of the Buddha. For instance, looking into a flower, you can see that the flower is made of many elements that we can call non-flower elements. When you touch the flower, you touch the cloud. You cannot remove the cloud from the flower, because if you could remove the cloud from the flower, the flower would collapse right away. You don't have to be a poet in order to see a cloud floating in the flower, but you know very well that without the clouds there would be no rain and no water for the flower to grow. So cloud is part of flower, and if you send the element cloud back to the sky, there will be no flower. Cloud is a non-flower element. ...if you continue, you will see a multitude of non-flower elements in the flower. In fact, a flower is made only with non-flower elements. It does not have a separate self.
>
> A flower cannot be by herself alone. A flower has to "inter-be" with everything else that is called non-flower. That is what we call inter-being. You cannot be, you can only inter-be.
>
> **Thich Nhat Hanh (1998)**

His Buddhist view of Life as "streams of being" and his views of all entities as non-separate but always "inter-being" fits with what we know of Life that is simultaneously organism, ecosystem, and biosphere, all enfolded together. We know this from the hard science of tracer studies just as clearly as we know it from Buddhist radical clarity of thought. No organism is separate from its surroundings. It is not simply an independent genetic code that prints *sui generis* an organism as if from a 3D printer. For example, a bear is not separate in any ultimate sense, as we saw in Jan Heath's beautiful glimpse of inter-being in Fig. 5.2. The bear we see and draw as bounded in space and time is really a stream of flies enfolded in fish, clouds, rains, and rivers enfolded into mountains, and oxygen, carbon dioxide, and other gases enfolded into wind and flowing masses of air across continents. The hyperset formulation is at work again across scales, embedding one another.

We humans are also in the process of inter-being. We can look inside ourselves and see ecosystems (such as the gut microbiome) and deeper still we see the environment (basic molecules and elements of physics and chemistry, energy quanta, and other building blocks). And, we can look outside and see manifestations of ourselves—the imprint of our ways of thinking and acting changes the world profoundly. To borrow another practice from Buddhism, meditation and a focus on one's breath, we have instant access to solid evidence of our unity with all Life. To breathe in and focus on one's inhaling breath is to be aware that invisible oxygen unites us with the green plants that are necessary extensions of every second of our living existence. To breathe out and focus on one's exhaling breath can foster awareness that invisible carbon dioxide is one way we return the favor and are in a process of mutually beneficial inter-being with green plants. If we want both the inner and outer realms to be beautiful, healthy, and balanced, then we can start by mindfulness and understanding how they are unified, interdependent, and intertwined across many orders of magnitude of space and time from milliseconds to millennia and microns to mountain ranges.

Another message of sustainability and holistic Life science is that place matters, and that investment in and commitment to place and the integration and stewardship of the local natural environment are essential to human personal and social well-being. Fig. 9.1 is a photograph of an old hillside town named Vranov nad Dyji, which is in the Czech Republic just across the Austrian border. The castle on the hill, overlooking the Dyji River (Thaya in German), dates from c. 1100,

FIGURE 9.1

Vranov nad Dyji, Czech Republic.

Photo credit B.D. Fath.

and the human edifices emerge out of, fit in well with, and complement the natural landscape. The area has been able to provide a sustainable, continuous home to humans for almost 1000 years. Possibly, we could say humans have improved the Life—environment system of this region and made the place better and more productive through their actions rather than consuming and exploiting it. Protecting and enhancing an integrated socio—ecological system returns us to a core value of love of life. We assume that the multigenerational denizens of Vranov nad Dyji love, respect, and take pride in their place. This is a place worth investing in, renewing, and defending, which it has surely had to do numerous times in its existence.

In fact, if one looks carefully, perhaps squinting with our mind's eye, we can see the overlapping and interconnected holons at various scales—organism, ecosystem, and biosphere—which give rise to the autocatalytic cycles that maintain the region's sustenance (Fig. 9.2). This is a living example of what we are aiming for in a win—win human and Life—environment relation and the opposite of many more recent places that lack distinction, life, value, or conviction, the so-called modern built environments referred to as "The Geography of Nowhere" (Kunstler, 1994). No wonder these places that belong nowhere are neglected and disabused, because they have no sense of space, no investment in lasting gradient formation, and no connective time thread woven through them. They are treated as a commodity where resources (including the space itself) are to be siphoned off until any value is gone and then prompting a move to the next place; lather, rinse, repeat, again and again—the very definition of unsustainability.

FIGURE 9.2

Conceptualization of the holons of Life or "flexons" at work in a real landscape. Organism, ecosystem, and environment hexaflexagon images from Sarah McManus.

While the real, unique, and specific details of local environment have been integrated in the Vranov nad Dyji landscape, the same technologies, applications, policies, rules, practices, infrastructure, and processes would not work if we attempted to transfer them directly to other local environments. Likewise, what has worked in the pictured landscape for a small population may not be scaled up to continental or global scales as is. Perhaps, we have already exceeded the global carrying capacity for this way of small-scale living. However, our challenge is to take the lessons learned so that they could be scaled up, generalized, abstracted, and then remodeled for unique application in each unique local context. Thus, we also need the universal and global principles of Life—environment and these must also be understood, aligned, and successfully integrated. So, we find yet another set of dialectics of local versus global, and real/specific versus abstract/generic, involved in learning how to live in a win—win relation with our Life-support context.

THE ROADS AHEAD, THE ROAD MAPS, AND THE DRIVERS

Yogi Berra, the major league baseball player famous for his oddball insights, once said "If you don't know where you are going, you will probably end up somewhere else." This is a humorous analogy to what we brilliant humans may inadvertently be doing with respect to our relationship to our environment, home, and collective futures. As we have worked to illustrate from multiple angles, unless we start, proceed always, stay focused, steer effectively, and strive for outcomes with clearly articulated Life value, we run the risk that we will end up somewhere else—we may not be able to continue the good Life we have come to assume and take for granted. We also see the added challenge of having the wisdom and clarity to know when we need to chart two distinct futures to match the fundamental natures of two distinct groups of people. If this seems like an act of splitting or fragmenting, then remember that we have always described the Sustainer and Transcender camps as necessarily complementary and both united in the larger unifying project of Life itself.

Possible rewordings of Berra's funny aphorism could be that "If the landscape has changed since the time your map was made, then by following it you may end up somewhere you did not intend to go." This fits with the story we have developed in which our science paradigm, economics, and other cultural "road maps" or systems of collective intelligence were developed at a different time in history, and the world has changed in fundamental ways since. Some of this change has to do with the scale and scope of the human enterprise with respect to the scale of the planetary environment, its material and energetic resources, and the real physical aspects of carrying capacity and Life support. Other aspects of the change have to do with what we have learned—under this heading, we must include eco-literacy (how Life is organized and functions), and the double-edged sword of over-using mechanistic models in science as our only road map.

A version of the Berra adage focused on two distinct groups and futures might be "If you have two drivers and they fight over which direction to go, you could wreck your car (or bike, or train), injure both drivers, and get nowhere." Given our current circumstances, which we suggest is analogous to this, it is important to stop, reflect, assess the situation, and find a path to peaceful resolution. Only after such resolution and clarification should we get back in the vehicle and proceed to move forward. Given our proposal for two distinct futures, this analogy requires either two separate vehicles for Sustainers and Transcenders or taking turns with one group driving and the other group enjoying the ride in supportive solidarity. The alternative car wreck or train wreck—a fitting analogy for our current crisis—should not be an option due to the obvious downside risk and permanent harm done.

We showed in Chapter 1 that our Sustainer/Transcender framework is compatible with the cultural theory distinction between Egalitarians and Individualists, which Thompson et al. (1990) viewed as groups with opposite emphasis on managing human needs versus managing natural resources, respectively. While this cultural classification scheme fits well with past and present human—environment conditions and relations, we think our categories are more forward compatible. This is because we see the Sustainer/Transcender dichotomy to be informative and relevant into the future and to help anticipate and understand a hypothetical coming bifurcation we foresee in human—environment cultural types. As the center of cultural intelligence, we again discuss the imperative that science instigate and lead this peaceful bifurcation revolution so that the two differentiated worldviews, systems of ideas, and action programs of Sustainers and Transcenders can differentiate while remaining in closely coordinated complementarity.

DISRUPTING THE SCIENTIFIC PARADIGM

The paradigm shift we propose with this book shares some similarities with "disruptive innovation" that has become more common in the business world in recent decades with business innovation like Netflix, Uber, Twitter, and other smaller firms that have upset dominant mainstream corporations in a given sector. Disruption in the business world was described well by Christensen et al. (2015).

> "Disruption" describes a process whereby a smaller company with fewer resources is able to successfully challenge established incumbent businesses. Specifically, as incumbents focus on improving their products and services for their most demanding (and usually most profitable) customers, they exceed the needs of some segments and ignore the needs of others. Entrants that prove disruptive begin by successfully targeting those overlooked segments, gaining a foothold by delivering more-suitable functionality—frequently at a lower price.

Incumbents, chasing higher profitability in more-demanding segments, tend not to respond vigorously. Entrants then move upmarket, delivering the performance that incumbents' mainstream customers require, while preserving the advantages that drove their early success. When mainstream customers start adopting the entrants' offerings in volume, disruption has occurred.

Instead of a smaller company, we see holistic Life science as a smaller system of ideas compared to the current mainstream science paradigm. We see the mainstream paradigm as "ignoring the needs" of the majority of people, and we are targeting those unfulfilled needs with the "more-suitable functionality" of holistic Life science including ecological network analysis, and the many other principles and methods we have demonstrated have unique capacity for understanding and problem solving beyond the reductionist, mechanistic mainstream. Similar to the passage above, we see the mainstream "not to respond vigorously"—excellent holistic and synthetic ideas and tools, sharing minority status with our views, have been widely ignored for decades. However, if we can "move upmarket" by "delivering the performance," we may in fact achieve disruptive innovation in the marketplace of scientific theories and practices.

In analogous fashion, Holling (e.g., Gunderson and Holling, 2002) presented and has continued to develop his synthetic idea of the adaptive cycle as a graphical means to understand growth, development, and disruption in natural ecosystems. In Fig. 9.3, we show a recent version of his adaptive cycle that describes the dominant process in an ecosystem as it proceeds through the four quadrants:

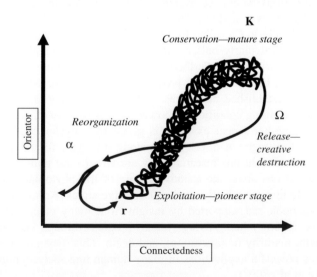

FIGURE 9.3

Modified version of Holling's four-stage adaptive cycle.

Reprinted from Burkhard, B., Fath, B.D., Müller, F., 2011. Adapting the adaptive cycle: hypotheses on the development of ecosystem properties and services. Ecol. Modell. 222, 2878–2890.

1. Bottom left, labeled r—This is the beginning of ecosystem development on a site, like early succession. This period is characterized by exploitation of natural resources and rapid growth of colonizer and pioneer species populations and biomass.
2. Top right, labeled K—This is late stage succession on a site, like mature forest, prairie, or other long-lasting and stable community-ecosystem. This period is characterized by conservation of natural resources and slow growth of late succession species populations and biomass.
3. Bottom right, labeled Ω- the end—This period, described as "release," is a period of disruption and reset. Catastrophic and systemic change can come from outside and be massive and widespread (e.g., fire, insect pest, or storm) or it can come from within and be small scale and piecemeal (e.g., deaths and destructive falls of very large trees creating forest gaps). This disruption releases some of the biomass and resources that had been tied up in the large stock of biomass and other Life—environment structures such as soils.
4. Middle left, labeled α- the beginning—This period, described as "reorganization," is the renewal and establishment of a novel community-ecosystem following the disruptive events of the release stage. The reorganization may lead to a system very similar to the previous one, or it may be very different based on contingencies and new conditions such as invading species, altered landscape or climate context, or other factors.

The conceptual framework of Holling's adaptive cycle has been applied successfully to numerous natural ecosystems. The ideas for this conceptual model first formed when he was studying for many years the spruce budworm outbreaks that occur periodically in Canadian forests (Holling, 1973). These outbreaks decimate the forests, particularly balsam fir. However, the outbreaks are rare, with only six since the early 1700s. Furthermore, without the outbreaks, the fir would dominate the forest, but the opening provides opportunity for other species of fir and spruce to be established and grow. In this manner, the ecosystem remains adaptive to the prevailing conditions following stages of growth, stasis, crisis, and reorganization. Another recent example is given by Angeler et al. (2015) who have quantified the stages of development along the cycle for Baltic Sea phytoplankton communities. In this research, they are able to use multivariate analysis to confirm predictions about the adaptive nature of the phytoplankton communities as they cycle through adaptive and conservative states. As we have expressed throughout this book and supported by insight from many others (eloquently by Wendell Berry, among others), living systems progress through inseparable stages of birth, growth, maturity (conservation), and death. This dynamic change seen in nature also has powerful applications in many human and socioeconomic systems as well (Folke et al., 2002).

The adaptive cycle has been applied to the growth and development of social institutions and firms (Fath et al., 2015) such that a resilient system is one that navigates the entire cycle successfully (Fig. 9.4). In addition to the r, K, Ω, and α, we add K_{lim}, which represents times during which crisis planning keeps the

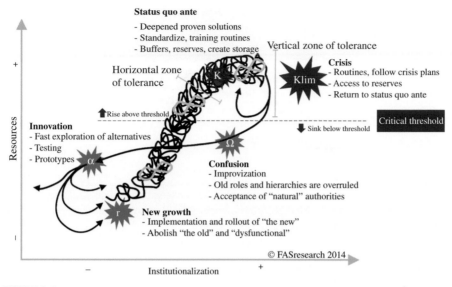

FIGURE 9.4

The adaptive cycle applied to the growth and development of social institutions and firms (Fath et al., 2015).

system above a critical threshold. However, one important message of viewing the system as a repetitive cycle is to anticipate and even welcome each stage as part of an interconnected whole. In other words, it is not just about always growing or always remaining on top, but being prepared and adaptive to changes that inevitably come. In particular, we addressed the pathologies that occur to keep the system stuck in one stage or inhibit it from entering another—two barriers that are not the same. For example, growth-oriented paradigms, policies, and practices—that debt-based money requires—favor the r-stage at the expense of reaching a conservative equilibrium in K-stage—there is a right time to grow but it is not always. The Ω-stage free-fall collapse will continue without redirection and reorientation in a new direction that could have come from contingency planning and capital accumulation during the previous K-stage.

Holling's approach to understanding systems dynamics in ecosystems shares generic similarities with the release, reorganization, and renewal that we see to be necessary in the science paradigm. The current dominant mainstream paradigm is much like a mature ecosystem—many resources are tied up in massive academic, government, and corporate structures, growth is very slow, certain highly competitive "species" of scientists reign atop the food chain, etc. An analog of a disruptive catastrophe from outside this science ecosystem would be global environmental catastrophe. Given the harm that this would cause, we seek to spark a release and reorganization from within so that we might be better prepared for and even prevent a global environmental catastrophe.

FIGURE 9.5

System dynamics from growth and development to conservation and stasis. Crisis can spur new ideas and stages. Modified from Goerner (2013), whose figure caption was "The Cycles of Civilization as Self-organizing Development."

Goerner (2013) contributed a similar conceptual approach with a focus on development within human culture and civilization. In Fig. 9.5, we show a modified version of her diagram for understanding crisis and Great Change in stages of civilization.

Goerner wrote of Fig. 9.5 (her Fig. 1):

> ENS [energy network science] places today's events within a natural development process called **self-organization**. In human systems, the standard self-organizing cycle of development proceeds as follows: 1) pent-up frustration leads to **pressure** for change; 2) some naturally-occurring **diversity**—an electrifying idea or a galvanizing leader—serves as a seed crystal around which momentum begins to organize; 3) some insightful group of individuals develops a **demonstrably better way** of doing business, economics, politics, education, etc., that provides a way out of the crisis; 4) the human system **reorganizes** in a way that relieves the pressure and advances societal development; 5) the new pattern of organization grows because energy, resources, and/ or information flows more robustly than before; 6) eventually the system reaches the limits of this pattern of organization, new pressure builds, and the pattern repeats leading to a new stage of development. The catch is that progress is not guaranteed; should any part of the process [be] suppressed, pressure will build into a snowballing series of calamities that may cause the society to regress or even collapse.

We see a very strong mapping of this generic scenario on our current crisis and when taking a radically empirical and critical hard look at mainstream science. The "pent-up frustration" comes from the failure of the current science enterprise to successfully address, understand, and solve the systemic environmental and cultural crises. This book and the extensive community of holistic and synthetic thinkers we have cited provide the "naturally occurring diversity," and we also believe many "electrifying ideas." As we continue to develop and apply holistic Life science, we are confident we will collectively realize a "demonstrably better way" of doing science and of living on Earth that simultaneously embraces both Sustainer and Transcender worldviews and action programs.

Note in Fig. 9.5 where a branch splits off to show two paths at the peak of the civilization development curve. We differ somewhat when applying Goerner's model to the present moment and the future where we foresee a coming bifurcation into complementary Sustainer and Transcender groups. Where Goerner shows "Regression or Collapse" after the peak, we would imagine something more like Odum and Odum's *Prosperous Way Down* and our own *Flourishing Within Limits*—a managing of human needs and a successful effort to live well within the real physical limits of the Earth that is the Sustainer future path.

But, it is also exciting and hopeful to examine the other path—the arrow upward that Goerner labeled "New Stage of Civilization." While we have to remember that both paths are unified and necessarily mutually aiding and interdependent, one aspect of the New Stage of Civilization must integrate the Transcender mode as well as the Sustainer mode. Whether transcending existing limits on Earth in nonmaterial progress, or by transcending Earth itself on the necessary project of helping Life advance into space, the deep and unstoppable Life urge of growth and innovation is just as important.

Brokering the peace accord between currently antagonistic Sustainer and Transcender tribes will not be easy, but it can be done. Transcenders might gain 50%−100% of all remaining fossil fuels, if they can agree that environmental Life support—atmosphere, soils, species and more—is sacred and inviolable for their new allies the Sustainers and their successful win−win future or Earth. This is just one rough example of concessions that could be imagined—but no such peace deals will be possible unless we reform our science to prepare the individual and collective intelligences that will come to the table at the summit.

We see a similar complex adaptive cycle like Holling's, and civilization development curve like Goerner's, needed for science so that it can participate in and lead the cultural bifurcation for two successful futures. This book, the many supporting references and authors we cite, and the community of holistic Life science leaders, can provide the "disruptive innovation" as an intentional paradigm shift. We need this shock in science to jump-start the process of waking up, clarifying values and thinking, leaping into action, and leading by example.

We see no alternative—science must lead. Hypocrisy and exceptionalism within academic, corporate, nonprofit, and government science is perhaps the single worst infection of our systemic disease. You need not take only our word for it. Aldo Leopold (Leopold, 1993) wrote as much:

> One of the penalties of an ecological education is that one lives alone in a world of wounds. Much of the damage inflicted on land is quite invisible to laymen. An ecologist must either harden his shell and make believe that the consequences of science are none of his business, or he must be the doctor

who sees the marks of death in a community that believes itself well and does not want to be told otherwise. (p. 165)

The consequences of science are our business. We must transform our externalizing of our own environmental impacts—like a form of claiming science is exceptional and thus allowed to cause environmental damage. While this mismatch of knowledge and action exists, while the center of our societal wisdom is confused and schizophrenic, we can have no hope that some other sector will do a better job, figure things out, and deliver solutions. As Thich Nhat Hanh might say, there is no way to sustainability; sustainability is the way. The means and ends must be unified, much as the many other fragmented and disjointed threads we have suggested weaving back together.

We must affirm Life value now, and we must protect and reinforce it heavily at every step in our scientific enterprises, so that the technology, applications, policy, and governance that diffuse out from and coexist with science result in win—win human—environmental relations. Reward structures in academia, as well as taxes, regulations, rituals, celebrations, monuments—all aspects of culture must collaborate and assist to value and sustain Life for it to improve rather than degrade due to lack of understanding, lack of care, and lack of attention. We can do science **and** improve the environment. Life has already shown us this is possible, and Life teaches us how to accomplish it. Science is one step in many that need reconsideration and transformation. From this, we envision other possibilities such as:

We can cultivate agriculture **and** improve the environment.

We can build cities **and** improve the environment.

We can be human **and** improve the environment.

Since science informs, guides, and supports all of these human endeavors and more, it has an especially great responsibility for self-consistent theory and action, principles and behavior, knowledge (science) and ethics (con-science as Thoreau referred to it).

THE ECOLOGICAL SELF AND THE PARTICIPATORY UNIVERSE

Another problem we face is that even with the understanding and motivation to change, inertia of societal forces can quickly impede and thwart actual change. The ideal conditions would entail a context in which action for change would be supported, rewarded, and amplified by society and economy. Two views of this ideal of spontaneous, organic, effortless, or frictionless change can help to imagine how we can create and foster such a supportive context for Great Change and sustainability.

Arne Næss's deep ecology and ecological self (1987, 1988, 1989) provide insight that aligns with both the Buddhist views of Thich Nhat Hanh and the mathematical hyperset formalism of Life—environment as unfractionable and unified whole. Næss's deep ecology is compatible with the principles in this book, the ecological metaphysic of Ulanowicz (1999a), and a holistic mental model of Life and self as extensive in space and continuous in time beyond the body and life span of the individual human organism. Næss (1987) wrote:

> The requisite care flows naturally if the 'self' is widened and deepened so that protection of free nature is felt and conceived as protection of ourselves.
>
> If reality is like it is experienced by the ecological self, our behaviour naturally and beautifully follows strict norms of environmental ethics. We certainly need to hear about our ethical shortcomings from time to time, but we change more easily through encouragement and through a deepened perception of reality and our own self—that is, a deepened realism. (p. 40)

Providing a "hard" scientific basis for these views, and promoting these widely through educational and media enterprises, should complement and help amplify the other evidence, intuitive appeal, and ethical power of Næss's "ecological self," deep ecology, and other work. This, in turn, could help with practical development of socioeconomic context so that actions like recycling, composting, using solar energy, or biking to work can be rewarded rather than penalized, and that they occur in the natural act of carrying out one's activities, process enabling process, resulting in closure to efficient cause. The linking of supporting processes will be the key to achieving a sustainable way.

The Story of Change is an excellent video produced by The Story of Stuff Project (storyofstuff.org/movies/story-of-change/). This animated video explains how unless mainstream socioeconomic "flow" changes to support self-change for sustainability, actual progress comes at a great cost and effort for the change agents and leaders who have to swim against a larger and stronger tide. They also suggest how policies and values can be aligned to aid rather than hinder correct action and emphasize the importance of engaged citizens who actively participate in the democratic governance process. To achieve a "level playing field" would be a step in the right direction. To tilt the playing field in favor of rewarding those who transform themselves and their organizations to renewable energy, recycling materials processes, ensuring environmental quality improvement, and related Life-valuing actions would be even better. Full accounting and internalizing all environmental and social costs would be an essential aspect of this correction of the tilt of the playing field.

These are examples of the influence of the ideas in this book on some of the day-to-day thoughts and actions we may experience as we grapple with our systemic humans-in-the-environment crisis. Another idea within current conventional wisdom that has diffused out from science is that entropy and the spontaneous trend toward disorder is the dominant trend in the behavior of all systems. This deeply ingrained idea is part of why we learned to accept environmental degradation as somehow in line with the necessary workings of the universe. We have already shown this to be false with respect to the many ways the Life has improved its environment, increased order, and built rather than dissipated gradients and stores of high quality energy. While much more abstract, far off, and thus the opposite of practical solutions for near-term improvement, a thought experiment may help us to challenge several more components of the societal working assumption of the dominance of the second law of thermodynamics.

One of the ways that biology and ecology textbooks explain Life's non-entropic capacities is by reference to the continual stream of high-quality energy from the Sun that fuels all Life activity. To extend this explanation to one logical extreme would be to accept that Life on Earth is forever dependent on incoming energy from the Sun. But, this also provides an interesting avenue by which to imagine a test of the dominant entropic tendency. If we could escape from the solar system—if we could successfully colonize Life beyond Earth in some space station or on some other planet so as to be free from any dependency on the Sun, then this would call into question if not outright falsify the hypothesis (now treated as law) that entropy is greater in influence than syntropy or the inherent self-organizing capacity of the universe.

Frances Moore Lappé and the Story of Change folks have demonstrated how participatory engagement in democratic society and governance is essential for societal transformation for sustainability. We maybe able to generalize and extend this radically to consider a participatory universe in which via science, technology, and culture we make predictions, set up experiments, test theories, and even have a participatory hand in determining the balance of syntropy and entropy tendencies in the universe.

CLARITY OF MIND AND PURPOSE

The conventional wisdom ideas of "looking out for number one" and "survival of the fittest" that have diffused outward from the neo-Darwinian scientific paradigm are in need of major update for a postmodern anthropocene world in which humans have filled and now dominate the planet. As Gregory Bateson pointed out, and as we quoted him earlier, any organism that operates under that narrow and selfish organismal principle will destroy its environment and then destroy itself. Given the science of networks, webs, and interdependent flows and processes, we have shown multiple times, and that can be found in even greater abundance and richer detail in the publications of systems and network ecologists, we know well that no individual self can be fragmented from its environment without fatal damage. Updated mottos that embody this relational wisdom would be:

> Looking out for #1 — where #1 is Life as a unified whole
> Survival of the fittest self-world and Life—environment relations
> Spoils go to the victor who promotes self-sustaining (autocatalytic) cycles

These principles are not self-sacrificing or naively selfless—ask any of the millions of species and billions of individual organisms on Earth, essentially all

Life outside of industrial culture. The same integrated ethics of self-improvement in the context of environmental home improvement can allow us to act for the good of ourselves and for our children, grandchildren, and future generations simultaneously. And, in this well-founded ethic and holistic science that balances the smaller self-interest of the individual with the interest of the larger ecological self, we can have the best of both worlds. As if material and physical improvement (atmosphere, soils, biodiversity, and more) is not great enough reward for holistic thought-action, we gain yet another reward—the spiritual sense of acting for something larger than and longer lasting than ourselves. This provides a team sense of purpose and togetherness which embraces not only all of humanity, but also all living beings and our shared living home.

We see humans as unique and crucial member species of Life as a whole, with unique capacities and important roles to play in service to Life. *Homo sapiens* is the only species able to develop the machines we will need to explore space and eventually colonize Life beyond Earth. We are the only species that can scan the skies for potential impactors and imagine scenarios to protect Earth against the catastrophic damage they could inflict. We are the only species able to understand, synthesize, and interpret knowledge of the entirety of Life—its full span in time and its full extent in space. We are the only species able to ponder and have a chance to comprehend Life "from origin to destiny." We are the only species able to harness the planetary gift and motherlode of fossil fuels—and we should use this power responsibly, and on behalf of Life itself. The list of our special skills, capacities, and responsibilities is very long and profound.

But we cannot achieve these great accomplishments while we remain confused and in conflict. In service to Life, let us begin the clarity and cooperation now.

Bibliography

AASHE STARS, 2018. Sustainability Tracking, Assessment and Rating System (STARS) of the Association for the Advancement of Sustainability in Higher Education. Available from: < https://stars.aashe.org/ >.

Aber, J.D., Kelly, T., Mallory, B. (Eds.), 2009. The Sustainable Learning Community: One University's Journey to the Future. University Press of England, Lebanon, NH, 267 pp.

Ackoff, R.L., 1974. Redesigning the Future: A Systems Approach to Societal Problems. Wiley, New York, NY.

ACUPCC, 2018. American College and University Presidents' Climate Commitment. Available from: < http://secondnature.org/climate-guidance/the-commitments/ >.

Agyeman, J. "Just Sustainabilities". Definition. Available from: < http://julianagyeman.com/ >. (Accessed on 2018).

Alexander, C., 1964. Notes on the Synthesis of Form. Harvard University Press, Cambridge, MA, 216 pp.

Alexander, C., 2012. The Battle for the Life and Beauty of the Earth: A Struggle between Two World-Systems. Oxford University Press, pp. 520.

Allen, T.F.H., Hoekstra, T.W., 1992. Toward a Unified Ecology. Columbia University Press, New York, NY, 384 pp.

Angeler, D.G., Allen, C.R., Garmestani, A.S., Gunderson, L.H., Hjerne, O., Winder, M., 2015. Quantifying the adaptive cycle. PLoS One 10 (12), e0146053. Available from: https://doi.org/10.1371/journal.pone.0146053.

Asimov, I., Shulman, J.A. (Eds.), 1988. Isaac Asimov's Book of Science and Nature Quotations. Blue Cliff Editions, Inc. Weidenfeld & Nicolson, New York, NY, Also online at https://archive.org/stream/BookOfScienceAndNatureQuotations-IsaacAsimov/asimov-nature-quotes_djvu.txt.

Barrett, G.W., Peles, J.D., Odum, E.P., 1997. Transcending processes and the levels of organization concept. BioScience 47 (8), 531−535. ISSN 0006-3568.

Bateson, G., 1988. Mind and Nature: A Necessary Unity. Bantam Books, New York, NY, 257 pp.

Benyus, J., 1997. Biomimicry: Innovation Inspired by Nature. William Morrow, New York, NY, 288 p. ISBN-13: 978-0688136918.

Berry, W., 1977. The Unsettling of America: Culture and agriculture. Counterpoint Press, Berkeley, CA, 234 pp.

Berry, W., 1981. Solving for Pattern. Chapter 9 in The Gift of Good Land: Further Essays Cultural & Agricultural (North Point Press, 1981). Originally published in the Rodale Press periodical The New Farm.

Berry, W., 2001. Life Is a Miracle: An Essay Against Modern Superstition. Counterpoint Press, Berkeley, CA, 176 pp. ISBN 9781582431413.

Bicchieri, C., Muldoon, R., 2014. "Social Norms," The Stanford Encyclopedia of Philosophy, Spring. ed. In: Edward N. Zalta (Ed.). < https://plato.stanford.edu/archives/spr2014/entries/social-norms/ >.

Billings, W.D., 1952. The environmental complex in relation to plant growth and distribution. Q. Rev. Biol. 27 (3), 251−265.

Biomimicry, 2018. Biomimicry Institute. Available from: < https://biomimicry.org/ >.

Blackburn, S., 2016. The Oxford Dictionary of Philosophy. Oxford University Press, Oxford, 531 pp.

Blersch, D., Kangas, P., 2003. Mid-Season Status of the Mars Desert Research Station's Wastewater Treatment System. Project report available from: < https://enst.umd.edu/sites/enst.umd.edu/files/_docs/greenhouse1.pdf > .

Bohm, D., 1995. Wholeness and the Implicate Order. Routledge, New York, NY (after original 1980).

Bohr, N., 1949. Discussion with Einstein on Epistemological Problems in Atomic Physics. In: Schilpp, P.A. (Ed.), Albert Einstein: Philosopher—Scientist, pp. 240. This passage is also reproduced in Niels Bohr Collected Works, 7, 1996, 339—381.

Bondavalli, C., Ulanowicz, R.E., 1999. Unexpected effects of predators upon their prey: the case of the American Alligator. Ecosystems 2 (1), 49—63.

Borrett, S.R., Salas, A.K., 2010. Evidence for resource homogenization in 50 trophic ecosystem networks. Ecol. Modell. 221 (13—14), 1710—1716.

Bostrom, N., Yudkowsky, E., 2014. The ethics of artificial intelligence. In: Frankish, K., Ramsey, W.M. (Eds.), The Cambridge Handbook of Artificial Intelligence. Cambridge University Press, Cambridge, pp. 316—334. 365 pp. ISBN-13: 978-0521691918.

Boulding, K.E., 1966. The economics of the coming spaceship earth. In: Jarrett, H. (Ed.), Environmental Quality in a Growing Economy. Resources for the Future/Johns Hopkins University Press, Baltimore, MD, pp. 3—14.

Burkhard, B., Fath, B.D., Müller, F., 2011. Adapting the adaptive cycle: hypotheses on the development of ecosystem properties and services. Ecol. Modell. 222, 2878—2890.

Byers, J., 1996. Brothers in Spirit: The Correspondence of Albert Schweitzer and William Larimer Mellon, Jr. Syracuse University Press, Syracuse, NY.

Cabrera, D., Mandel, J.T., Andras, J.P., Nydam, M.L., 2008. What is the crisis? Defining and prioritizing the world's most pressing problems. Front. Ecol. Environ. 6, 469—475. ISSN 1540-9295.

Calaprice, A., 2005. The New Quotable Einstein. Princeton University Press, Princeton, NJ, ISBN: 0691120749.

Campbell, N.A., Reece, J.B., Urry, L.A., Cain, M.L., Wasserman, S.A., Minorsky, P.V., et al., 2008. Biology. Pearson/Benjamin Cummings, London, 1267 pp. ISBN: 9780805368444.

Capra, F., Luisi, P.L., 2014. The Systems View of Life: A Unifying Vision. Cambridge University Press, Delhi, 498 pp.

Chivian, D., Brodie, E.L., Alm, E.J., Culley, D.E., Dehal, P.S., DeSantis, T.Z., et al., 2008. Environmental genomics reveals a single-species ecosystem deep within Earth. Science 322 (5899), 275—278.

Christensen, C.M., Raynor, M.E., McDonald, R., 2015. What is disruptive innovation? Harvard Business Review. Available from: < https://hbr.org/2015/12/what-is-disruptive-innovation > .

Clark, A., Chalmers, D.J., 1998. The extended mind. Analysis 58, 7—19.

Covey, S.R., 1989. The Seven Habits of Highly Effective People: Restoring the Character Ethic. Simon and Shuster, New York, NY.

Darwin, C.R., 1859. On the Origin of Species by Means of Natural Selection, or the Preservation of Favoured Races in the Struggle for Life. John Murray, London.

Donne, J., 1997. Devotions Upon Emergent Occasions & Death's Duel. The Ages Digital Library, Albany, OR. Available online: http://media.sabda.org/alkitab-10/LIBRARY/INSPIRAT/DON_DEVO.PDF.

Eigen, M., 1995. What will endure of 20th century biology? In: Murphy, M.P., O'Neill, L. A.J. (Eds.), What Is Life? The Next Fifty Years: Speculations on the Future of Biology. Cambridge University Press, Cambridge, 204 pp. ISBN-13: 978-0521599399.

Elbow, P., 1986. Embracing Contraries: Explorations in Learning and Teaching. Oxford University Press, New York, NY, 314 pp.

Eliot, C.W. (Ed.), 1914. William Shakespeare (1564–1616). The Tragedy of Hamlet Prince of Denmark. The Harvard Classics. 1909–14. P.F. Collier & Son, New York. Published online by Bartleby.com, Inc. 2001. Available from: < http://www.bartleby.com/46/2/43.html > .

Elton, C.S., 1958. Ecology of Invasions by Animals and Plants. Chapman & Hall, London.

Fang, D., Chen, B., 2015. Ecological network analysis for a virtual water network. Environ. Sci. Technol. 49 (11), 6722–6730.

Fantappiè, L., 1942. Teoria Unitaria del Mondo Fisico e Biologico (The Unified Theory of the Physical and Biological World). Di Renzo Editore, Roma, 1991.

Fath, B.D., 2004. Distributed control in ecological networks. Ecol. Modell. 179, 235–246.

Fath, B.D., 2007. Network mutualism: positive community level relations in ecosystems. Ecol. Modell. 208, 56–67.

Fath, B.D., 2014. Sustainable systems promote wholeness-extending transformations: the contributions of systems thinking. Ecol. Modell. 293, 42–48.

Fath, B.D., 2015. Quantifying economic and ecological sustainability. Ocean Coast. Manage. 108, 13–19.

Fath, B.D., 2017. Systems ecology, energy networks, and a path to sustainability. Prigogine Lecture. Int. J. Ecodyn. 12 (1), 1–15.

Fath, B.D., Mueller, F., 2018. Conbiota. In: Fath, B.D. (Ed.), Encyclopedia of Ecology, second ed. Elsevier.

Fath, B.D., Patten, B.C., 1998. Network synergism: emergence of positive relations in ecological systems. Ecol. Modell. 107, 127–143.

Fath, B.D., Patten, B.C., 1999. Review of the foundations of network environ analysis. Ecosystems 2, 167–179.

Fath, B.D., Patten, B.C., Choi, J.S., 2001. Complementarity of ecological goal functions. J. Theor. Biol. 208, 493–506.

Fath, B.D., Scharler, U., Ulanowicz, R.E., Hannon, B., 2007. Ecological network analysis: network construction. Ecol. Modell. 208, 49–55.

Fath, B.D., Dean, C.A., Katzmair, H., 2015. Navigating the adaptive cycle: an approach to managing the resilience of social systems. Ecol. Soc. 20 (2), 24.

Fiscus, D.A., 2001–2002. The ecosystemic life hypothesis. In: Bulletin of the Ecological Society of America (three parts) Oct. 2001, Jan. and Apr. 2002.

Fiscus, D.A., 2007. Comparative Ecological Modeling for Long-term Solution of Excess Nitrogen Loading to Surface Waters and Related Chronic and Systemic Human-Environment Problems. Dissertation published in partial fulfillment of requirements of the Doctor of Philosophy degree in the University of Maryland Marine, Estuarine and Environmental Science Program. Appalachian Laboratory, Frostburg, MD.

Fiscus, D.A., 2013. Life, money and the "deep tangled roots" of systemic change for sustainability. World Fut. 69 (7–8), 555–571. Available from: http://www.tandfonline.com/eprint/JH4fJu8teEITj74yaujW/full.

Fiscus, D.A., Fath, B.D., Goerner, S., 2012. The tri-modal nature of life with implications for actualizing human-environmental sustainability. Emerg. Compl. Organ. 14 (3), 44–88.

Fiscus, D.A., Fath, B.D., Goerner, S.J., Ulanowicz, R.E., 2015. Testing applicability of an indicator of ecological network sustainability to socioeconomic systems. In: Presentation at the Ecological Society of America Conference. Aug. 14, 2015, Baltimore, MD.

Folke, C., Carpenter, S., Elmqvist, T., Gunderson, L., Holling, C.S., Walker, B., 2002. Resilience and sustainable development: building adaptive capacity in a world of transformations. AMBIO 31 (5), 437–440. Available from: https://doi.org/10.1579/0044-7447-31.5.437.

Fuller, R.B., 1970. Operating Manual for Spaceship Earth. Southern Illinois University Press, Carbondale, IL, 144 pp. ISBN-13: 978-0809303571.

Fuller, R.B., 1979. Synergetics 2: Explorations in the Geometry of Thinking. MacMillan Publishing Co, New York, NY.

Funtowicz, S.O., Ravetz, J.R., 1993. Science for the post-normal age. Futures 25, 739–755.

Giampietro, M., 2004. Multi-scale Integrated Analysis of Agroecosystems. CRC Press, Boca Raton, FL.

Goerner, S.J., 1999. After the Clockwork Universe: The Emerging Science and Culture of Integral Society. Triangle Center for Complex Systems, Chapel Hill, NC, 452 pp. ISBN: 9780863152900.

Goerner, S.J., 2013. Corrective lenses: how the laws of energy networks improve our economic vision. World Fut. 69 (7–8), 402–449.

Goerner, S.J., Dyck, R.G., Lagerroos, D., 2008. The New Science of Sustainability: Building a Foundation for Great Change. Triangle Center for Complex Systems, Chapel Hill, NC, 412 pp.

Goerner, S.J., Lietaer, B., Ulanowicz, R.E., 2009. Quantifying economic sustainability: implications for free-enterprise theory, policy and practice. Ecol. Econ. 69, 76–81.

Goerner, S.J., Fath, B.D., Fiscus, D.A., Ulanowicz, R.E., Berea, A., 2018. Measuring regenerative vitality: 10 systems principles for a healthy economy. In: Submitted to Ecology and Society.

Goleman, D., 2015. A Force for Good: The Dalai Lama's Vision for Our World. Bantam Books, New York, NY.

Goodland, R., Daly, H., 1996. Environmental sustainability: universal and nonnegotiable. Ecol. Appl. 6 (4), 1002–1017.

GPI, 2018. Genuine Progress Indicator. Available from: < http://rprogress.org/sustainability_indicators/genuine_progress_indicator.htm > .

Grossinger, R. (Ed.), 1978. Ecology and Consciousness. North Atlantic Books, Richmond, CA, 223 pp.

Guinée, J.B., 2006. Handbook on Life Cycle Assessment: Operational Guide to the ISO Standards. Springer Science & Business Media, Berlin, 692 pages.

Gunderson, L.H., Holling, C.S. (Eds.), 2002. Panarchy: Understanding Transformations in Human and Natural Systems. Island Press, Washington, DC, 536 pp. ISBN 9781559638579.

Hagens, N., 2018. Where Are We Going? The 40 Shades of Gray. Essay on the MAHB Blog. Available from < https://mahb.stanford.edu/blog/where-are-we-going/ > .

Hall, C.A.S., 2017. Energy Return on Investment: A Unifying Principal for Biology, Economics, Sustainability, XII. Springer International Publishing, Cham, 174 pp. ISBN: 978-3-319-47821-0.

Hanh, T.N., 1998. The Island of Self; The Three Dharma Seals. Dharma Talk. Available from < http://www.purifymind.com/IslandSelf.htm > .

Hardin, G., 1968. Tragedy of the commons. Science 162 (3859), 1243−1248.

Harding, S., 2006. Animate Earth: Science, Intuition and Gaia. Chelsea Green, Hartford, VT, 256 pp. ISBN: 9781933392295.

Henderson, L.J., 1913. The Fitness of the Environment: An Inquiry into the Biological Significance of the Properties of Matter. MacMillan Company, New York, NY.

Hetherington, J., 2018. Can robotic bees replace the real thing? Walmart files for patent for 'pollination drone'. Newsweek. March 15, 2108. Available from: < http://www.news-week.com/can-robotic-bees-replace-real-thing-walmart-files-patent-pollination-drone-845861 > .

Higashi, M., Patten, B.C., 1989. Dominance of indirect causality in ecosystems. Am. Nat. 133 (2), 288−302.

Holling, C.S., 1973. Resilience and stability of ecological systems. Annu. Rev. Ecol. Syst. 4, 1−23.

Hornborg, A., 2001. The Power of the Machine. Global Inequities of Economy, Technology and Environment. Altamira Press, Walnut Creek, CA.

Hornborg, A., 2011. Global Ecology and Unequal Exchange: Fetishism in a Zero-sum World. Routledge, Oxon, UK, pp. 208.

Hubbert, M.K., 1962. Energy Resources: A Report to the Committee on Natural Resources. National Academy of Sciences. National Research Council, ISBN: 978-0-309-35515-5. 153 pp. Available online at: http://nap.edu/21066.

Hubbert, M.K., 1976. Exponential growth as a transient phenomenon in human history. In: Paper presented before World Wildlife Fund, Fourth International Congress. San Francisco, CA, USA.

Hyslop, M., 2012. Health of Chesapeake Bay Gets Mixed Reviews: Crab Population Up; Overall Health Down. Gazette.net. Available from: < http://www.gazette.net/article/20120420/NEWS/704209687/1007/health-of-chesapeake-bay-gets-mixedreviews &template = gazette > .

IAN, 2012. Integration and Application Network Chesapeake Bay Report Card. Available from: < http://ian.umces.edu/ecocheck/report-cards/chesapeake-bay/2011/ > .

Ireland, K.A., 2010. Visualizing Human Biology. Wiley, Hoboken, NJ, 704 pp. ISBN: 9780470569191.

Jacobs, J., 1969. The Economy of Cities. Vintage Press, Random House, New York, NY, p. 268.

Jacobs, J., 2000. The Nature of Economies. Modern Library, New York, NY, 208 pp. ISBN-13: 978-0679603405.

Jobbagy, E.G., Jackson, R.B., 2000. The vertical distribution of soil organic carbon and its relation to climate and vegetation. Ecol. Appl. 10 (2), 423−436.

Jobbagy, E.G., Jackson, R.B., 2001. The distribution of soil nutrients with depth: global patterns and the imprint of plants. Biogeochemistry 53, 51−77.

Jørgensen, S.E., Fath, B.D., Bastianoni, S., Marques, J.C., Müller, F., Nielsen, S.N., et al., 2007. Systems Ecology: A New Perspective. Elsevier, Amsterdam, pp. 288.

Jørgensen, S.E., Fath, B.D., Nielsen, S.N., Pulselli, F.M., Fiscus, D.A., Bastianoni, S., 2015. Flourishing Within Limits to Growth: Following Nature's Way. EarthScan/Routledge, London, 236 pp.

Kangas, P., 2004. Ecological Engineering: Principles and Practices. CRC Press, Boca Raton, FL, pp. 469.

Kauffman, S., 1995. At Home in the Universe: The Search for the Laws of Self-organization and Complexity. Oxford University Press, New York, NY, 321 pp.

Kauffman, S., 2000. Investigations. Oxford University Press, Oxford, 302 pp. ISBN-13: 978-0195121049.

Kauffman, S., 2011. The End of a Physics Worldview: Heraclitus and the Watershed of Life. Cosmos and Culture. National Public Radio Blog. Available from: < www.npr.org/sections/13.7/2011/08/08/139006531/the-end-of-a-physics-worldview-heraclitus-and-the-watershed-of-life > .

Keller, E.A., Botkin, D.B., 2008. Essential Environmental Science. Wiley, Hoboken, NJ, 480 pp. ISBN-13: 978-0471704119.

Kercel, S.W., 2003. Endogeny and Impredicativity. In: IEEE Conference Publication. Obtained from the author via electronic mail.

Kercel, S.W., 2007. Entailment of ambiguity. Chem. Biodivers. 4 (10), 2369−2385.

Kharrazi, A., Rovenskaya, E., Fath, B.D., Yarime, M., Kraines, S., 2013. Quantifying the sustainability of economic resource networks: an ecological information-based approach. Ecol. Econ. 90, 177−186.

Kiernan, T. (Ed.), 1965. A Treasury of Albert Schweitzer. The Citadel Press, New York, NY, 349 pp.

Kirmayer, L.J., Dandeneau, S., Marshall, E., Phillips, M.K., Williamson, K.J., 2011. Rethinking resilience from indigenous perspectives. Can. J. Psych. 56 (2), 84−91.

Koestler, A., 1968. Ghost in the Machine: The Urge to Self-Destruction, a Psychological and Evolutionary Study of Modern Man's Predicament. Macmillan Publishers, London, 384 pp. ISBN: 9780090838806.

Kuhn, T.S., 1962. The Structure of Scientific Revolutions. University of Chicago Press, Chicago, IL, 264 pp. (1996), ISBN: 9780226458083.

Kunstler, J.H., 1994. The Geography of Nowhere: The Rise and Decline of America's Man-Made Landscape. Simon and Schuster, New York, NY, pp. 303.

Lahav, N., 1999. Biogenesis: Theories of Life's Origin. Oxford University Press, Oxford, 368 pp. ISBN: 9780195117554.

Lappé, F.M., 2011. EcoMind: Changing the Way We Think, to Create the World We Want. Nation Books, Small Planet Institute, New York, NY, 288 pp.

Leigh, P., 2005. The ecological crisis, the human condition, and community-based restoration as an instrument for its cure. Ethics Sci. Environ. Polit. 2005, 3−15. ISSN 1611-8014.

Leopold, A., 1949. A Sand County Almanac: And Sketches Here and There. Oxford University Press, New York, NY, 240 pp.

Leopold, L.B. (Ed.), 1993. Round River: From the Journals of Aldo Leopold. Oxford University Press, New York, NY.

Lewis, C.S., 1943. The Abolition of Man. Oxford University Press, London.

Lietaer, B., Arnsperger, C., Goerner, S., Brunnhuber, S., 2012. Money and Sustainability: The Missing Link. Triarchy Press, Devon.

Likens, G.E., Bormann, F.H., Johnson, N.M., Fisher, D.W., Pierce, R.S., 1970. Effects of forest cutting and herbicide treatment on nutrient budgets in the Hubbard Brook watershed-ecosystem. Ecol. Monogr. 40 (1), 23−47.

Lindeman, R.L., 1942. The trophic-dynamic aspect of ecology. Ecology 23, 399−418.

Lorenz, E.N., 1963. Deterministic nonperiodic flow. J. Atmos. Sci. 20, 130−141.

Lotka, A.J., 1925. Elements of Physical Biology. Williams and Wilkins Company, Baltimore, MD, 495 pp. (2011), ISBN: 9781178508116.

Louie, A.H., Poli, R., 2011. The spread of hierarchical cycles. Int. J. Gen. Syst. 40 (3), 237–261.

Lovelock, J.E., 1972. Gaia as seen through the atmosphere. Atmos. Environ. 6, 579–580.

Lovelock, J.E., 1988. The Ages of Gaia: A Biography of Our Living Earth. Oxford University Press, Oxford, 278 pp. ISBN: 9780393312393.

Lovelock, J.E., Margulis, L., 1974. Atmospheric homeostasis by and for the biosphere: the Gaia hypothesis. Tellus 26 (1–2), 2–10.

MacArthur, R.H., 1955. Fluctuations of animal populations and a measure of community stability. Ecology 36, 533–536.

MacFarquhar, L., 2018. Mind Expander. The New Yorker. April 2, 2018 Issue. Available from: < https://www.newyorker.com/magazine/2018/04/02/the-mind-expanding-ideas-of-andy-clark > .

Macy, J., 1991. Mutual Causality in Buddhism and General Systems Theory: The Dharma of Natural Systems. State University of New York Press, Albany, NY, 236 pp.

MBDC, 2018. McDonough Braungart Design Chemistry. Available from: < www.mbdc.com > .

McDonough, W., Braungart, M., 2002. Cradle to Cradle: Remaking the Way We Make Things. North Point Press, New York, NY, 196 pp.

McManus, S., 2013. Hexaflexagon Design for Representing the Hyperset Formalism Life. Unpublished work sent by email.

Meadows, D., 1999. Leverage Points: Places to Intervene in a System. Sustainability Institute. Hartland, VT, USA. Available from: < http://donellameadows.org/archives/leverage-points-places-to-intervene-in-a-system/ > .

Meadows, D.H., 2001. Dancing With Systems. Whole Earth. Available from: < http://www.wholeearth.com/issue/2106/article/2/dancing.with.systems > .

Mellor, M., 2010. The Future of Money: From Financial Crisis to Public Resource. Pluto Press, London, 208 pp. ISBN-13: 978-0745329956.

Mikulecky, D.C., 1999. Robert Rosen: The Well Posed Question and Its Answer — Why Are Organisms Different From Machines? Available from: < http://www.people.vcu.edu/~mikuleck/ > .

Millennium Ecosystem Assessment, 2005. Ecosystems and Human Well-being: Synthesis. World Resources Institute and Island Press, Washington, DC. Available online: https://www.millenniumassessment.org/documents/document.356.aspx.pdf.

Moeller, H., 2006. Luhmann Explained: From Souls to Systems. Open Court, Chicago, IL, 312 pp. ISBN-13: 978-0812695984.

Mollison, B., 1996. Permaculture: A Designer's Manual. Tagari Publications, Tyalgum.

Morowitz, H.J., 1992. The Beginnings of Cellular Life: Metabolism Recapitulates Biogenesis. Yale University Press, New Haven, CT, ISBN: 9780300102109.

Muir, J., 1911. My First Summer in the Sierra. Houghton Mifflin Company, Boston, MA, and New York, NY. The Riverside Press Cambridge. Quote from page 110. See also Sierra Club Muir exhibit at < https://vault.sierraclub.org/john_muir_exhibit/ > Also see full paragraph of the quote below (end of references).

Murphy, M.P., O'Neill, L.A. (Eds.), 1995. What Is Life? The Next Fifty Years: Speculations on the Future of Biology. Cambridge University Press, Cambridge, 204 pp. ISBN-13: 978-0521599399.

Næss, A., 1987. Self-realization: an ecological approach to being in the world. The Trumpeter: J. Ecosophy 4, 35–42.

Næss, A., 1988. Deep ecology and ultimate premises. Ecologist 18, 128−131.

Næss, A., 1989. Ecology, Community, and Lifestyle: Outline of an Ecosophy. (D. Rothenberg, Trans. and revised). Cambridge University Press, Cambridge.

NASA Astrobiology, 2017. About Astrobiology. Available from: < https://astrobiology. nasa.gov/about/ > .

NASA OSS, 2017. Mission of NASA Space Science. Available from: < https://www.hq. nasa.gov/office/nsp/oss.htm > .

Nash, R.F., 1989. The Rights of Nature: A History of Environmental Ethics. The University of Wisconsin Press, Madison, WI, 290 pp.

Neal, E.C., Patten, B.C., DePoe, C.E., 1967. Periphyton growth on artificial substrates in a radioactively contaminated lake. Ecology 48, 918−924.

Nicolis, G., Prigogine, I., 1977. Self-organization in Nonequilibrium Systems: From Dissipative Structures to Order Through Fluctuations. Wiley-Interscience, New York, NY, 512 pp. ISBN-13: 978-0471024019.

North, E.W., Houde, E.D., 2001. Retention of white perch and striped bass larvae: biological-physical interactions in Chesapeake Bay estuarine turbidity maximum. Estuaries 24 (5), 756−769.

Odum, E.P., 1969. The strategy of ecosystem development. Science 164, 262−270.

Odum, H.T., 1971. Environment, Power and Society. Wiley-Interscience, New York, NY, 331 pp. ISBN: 9780471652755.

Odum, H.T., Odum, E.C., 2001. A Prosperous Way Down: Principles and Policies. University Press of Colorado, Louisville, CO, 344 pp.

Olomucki, M., 1993. The Chemistry of Life. McGraw-Hill, New York, NY, USA, pp. 132. ISBN: 9780070479296.

O'Neill, R.V., DeAngelis, D.L., Waide, J.B., Allen, T.F.H., 1987. A Hierarchical Concept of Ecosystems. Princeton University Press, Princeton, NJ, pp. 253.

Orr, D., 1991. What Is Education For? Six Myths About the Foundations of Modern Education, and Six New Principles to Replace Them. In Context. Winter. Available from: < https://www.context.org/iclib/ic27/orr/ > .

Orr, D., 2008. Ecological systems thinking. In: Jørgensen, S.E., Fath, B.D. (Eds.), Encyclopedia of Ecology. Elsevier, Oxford, pp. 1117−1121.

Palmer, P.J., Zajonc, A., Scribner, M., 2010. The Heart of Higher Education: A Call to Renewal. Jossey-Bass, an imprint of Wiley, San Francisco, CA, 256 pp.

Patten, B.C., 1978. Systems approach to the concept of the environment. Ohio J. Sci. 78, 206−222.

Patten, B.C., 1981. Environs: the superniches of ecosystems. Amer. Zool. 21, 845−852.

Patten, B.C., 1982. Environs: relativistic elementary particles for ecology. Am. Nat. 119 (2), 179−219.

Patten, B.C., 1983. On the quantitative dominance of indirect effects in ecosystems. In: Lauenroth, W.K., Skogerboe, G.V. (Eds.), Analysis of Ecological Systems: State-of-the-art in Ecological Modelling. Elsevier, Amsterdam, pp. 27−37. [Presents the oyster reef model].

Patten, B.C., 1991. Network ecology: indirect determination of the life-environment relationship in ecosystems. In: Higashi, M., Burns, T. (Eds.), Theoretical Studies of Ecosystems: The Network Perspective. Cambridge University Press, Cambridge, pp. 288−351. ISBN: 9780521361385.

Patten, B.C., 1994. Ecological systems engineering: toward integrated management of natural and human complexity in the ecosphere. Ecol. Modell. 75/76, 653−665.

Patten, B.C., 1999. Out of the clockworks. Estuaries 22, 339−342.

Patten, B.C., 2001. Jacob von Uexküll and the theory of environs. Semiotica 134 (1/4), 423−443.

Patten, B.C., 2014. Systems ecology and environmentalism: getting the science right. Part I: facets for a more holistic *Nature Book* of ecology. Ecol. Modell. 293, 4−21.

Patten, B.C., 2016a. The cardinal hypotheses of *Holoecology*: facets for a general systems theory of the organism−environment relationship. Ecol. Modell. 319, 63−111.

Patten, B.C., 2016b. Systems ecology and environmentalism: getting the science right. Part II: the *Janus Enigma Hypothesis*. Ecol. Modell. 335, 101−138.

Patten, B.C., Witkamp, M., 1967. Systems analysis of 134cesium kinetics in terrestrial microcosms. Ecology 48, 813−824.

Patten, B.C., Bosserman, R.W., Finn, J.T., Cale, W.G., 1976. Propagation of cause in ecosystems. In: Patten, B.C. (Ed.), Systems Analysis and Simulation in Ecology, Vol. 4. Academic Press, New York, NY, pp. 457−579.

Pepper, S., 1942. World Hypotheses: Prolegomena to Systematic Philosophy and a Complete Survey of Metaphysics. University of California Press, Berkeley, CA, 348 pp.

Raatikainen, P., 2018. Gödel's Incompleteness Theorems. The Stanford Encyclopedia of Philosophy, Summer 2018 ed. In: Zalta, E.N. (Ed.). Available from: < https://plato.stanford.edu/archives/sum2018/entries/goedel-incompleteness/ > .

Raskin, P., 2017. Journey to Earthland: The Great Transition to Planetary Civilization. Tellus Institute, Boston, MA.

Raynaud, X., Nunan, N., 2014. Spatial ecology of bacteria at the microscale in soil. PLoS One . Available from: https://doi.org/10.1371/journal.pone.0087217.

Rockström, J., Steffen, W., Noone, K., Persson, Å., Chapin III, F.S., Lambin, E.F., et al., 2009. A safe operating space for humanity. Nature 461, 472−475.

Rosen, R., 1958. A relational theory of biological systems. Bull. Math. Biophys. 20, 245−260.

Rosen, R., 1977. Complexity as a system property. Int. J. Gen. Syst. 3 (4), 227−232.

Rosen, R., 1985. Anticipatory Systems: Philosophical, Mathematical and Methodological Foundations. Pergamon Press, Oxford, 441 pp. ISBN: 9780080311586.

Rosen, R., 1991. Life Itself: A Comprehensive Inquiry into the Nature, Origin, and Fabrication of Life. Columbia University Press, New York, NY, 285 pp. ISBN: 9780231075657.

Rosen, R., 2000. Essays on Life Itself. Columbia University Press, New York, NY, 416 pp. ISBN: 9780231105118.

Rosental, S. (Ed.), 1967. Niels Bohr: His Life and Work as Seen by His Friends and Colleagues. Interscience (Wiley), New York, NY, 355 pp.

Rowe, J.S., 1961. The level-of-integration concept and ecology. Ecology 42 (2), 420−427.

Saffo, P., 2013. The coming fight between engineers and druids. Edge . Available online at https://www.edge.org/response-detail/23858.

Salas, A.K., Borrett, S.R., 2011. Evidence for the dominance of indirect effects in 50 trophic ecosystem networks. Ecol. Modell. 222, 1192−1204.

Salthe, S., 1985. Evolving Hierarchical Systems. Columbia University Press, New York, NY, 343 pp. ISBN-13: 978-0231060165.

Salthe, S.N., 2001. Theoretical biology as an anticipatory text: the relevance of Uexkull to current issues in evolutionary systems. Semiotica 134 (1/4), 359−380.

Schneider, E.D., Sagan, D., 2005. Into the Cool: Energy Flow, Thermodynamics and Life. University of Chicago Press, Chicago, IL, 378 pp. ISBN-13: 978-0226739373.

Schrödinger, E., 1944. What Is Life? The Physical Aspect of the Living Cell. Cambridge University Press, Cambridge.

Schroeder, M., 2016. "Value Theory," The Stanford Encyclopedia of Philosophy, Fall ed. In: Zalta, E.N. (Ed.). Available from: < https://plato.stanford.edu/archives/fall2016/entries/value-theory/ > .

Schweitzer, A., 1965. The Teaching of Reverence for Life. Holt, Rinehart and Winston, New York, NY, 63 pp.

Schweitzer, A., 1969. Reverence for Life. Harper and Row, New York, NY, 153 pp.

Shani, I., 2014. Naturalized sacredness? A realist, panentheist, and perennialist alternative to Kauffman's constructivism. Zygon 49 (1), 22−41.

Slaper, H., Velders, G.J., Daniel, J.S., de Gruijl, F.R., van der Leun, J.C., 1996. Estimates of ozone depletion and skin cancer incidence to examine the Vienna Convention achievements. Nature 384 (6606), 256−259.

Slaper, T.F., Hall, T.J., 2011. The triple bottom line: what is it and how does it work? Indiana Bus. Rev. 86 (1), 4−8.

Soil Health Institute, 2017. Available from: < http://soilhealthinstitute.org/national-soil-health-measurements-accelerate-agricultural-transformation/ > .

Solomonoff, R.J., 1997. The discovery of algorithmic probability. J. Comput. Syst. Sci. 55 (1), 73−88.

Swenson, R., 1989. Emergent evolution and the global attractor: the evolutionary epistemology of entropy production. In: Proceedings of the 33rd Annual Meeting of the International Society for the Systems Sciences, vol. 3. pp. 46−53.

Szent-Gyorgyi, A., 1977. Drive in living matter to perfect itself. Synthesis 1 (1), 14−26.

Tansley, A.G., 1935. The use and abuse of vegetational terms and concepts. Ecology 16 (3), 284−307. Available from: https://doi.org/10.2307/1930070.

The Carbon Underground, 2018. Regenerative agriculture defined. White paper available from: < https://thecarbonunderground.org/wp-content/uploads/2017/02/Regen-Ag-Definition-7.27.17-1.pdf > .

Thompson, M., Ellis, R., Wildavsky, A., 1990. Cultural Theory. Westview Press, Boulder, CO.

Thoreau, H.D., 1854. Walden; Or, Life in the Woods. Ticknor and Fields, Boston, MA.

Town Creek Foundation, 2018. Website With Information on Chesapeake Bay Watershed and Its Systemic Environmental Problems. Available from: < https://www.towncreekfdn.org/ > .

Uhl, C., 2003. Developing Ecological Consciousness: Path to a Sustainable World. Rowman and Littlefield Publishers, Inc, Lanham, MD, 378 pp.

Uhl, C., 2013. Developing Ecological Consciousness: Path to a Sustainable World. Rowman & Littlefield Publishers, Lanham, MD, 379 pp.

Uhl, C., Stuchul, D.L., 2011. Teaching as If Life Matters: The Promise of a New Education Culture. Johns Hopkins Press, Baltimore, MD, 224 pp. ISBN-13: 978-1421400396.

Uhl, C., Kulakowski, D., Gerwing, J., Brown, M., Cochrane, M., 1996. Sustainability: a touchstone concept for university operations, education, and research. Conserv. Biol. 10, 1308−1311.

Ulanowicz, R.E., 1980. An hypothesis on the development of natural communities. J. Theor. Biol. 85, 223−245.

Ulanowicz, R.E., 1997. Ecology: The Ascendent Perspective. Columbia University Press, New York, NY, 201 pp. ISBN: 9780231108294.

Ulanowicz, R.E., 1999a. Life after Newton: an ecological metaphysic. BioSystems 50, 127−142. ISSN 0303-2647.

Ulanowicz, R.E., 1999b. Out of the clockworks: a response. Estuaries 22, 342−343.

Ulanowicz, R.E., 2001. The organic in ecology. Ludus Vitalis 9 (15), 183−204.

Ulanowicz, R.E., 2002. Ecology, a dialogue between the quick and the dead. Emergence 4, 34−52.

Ulanowicz, R.E., 2004. Quantitative methods for ecological network analysis. Comput. Biol. Chem. 28, 321−339.

Ulanowicz, R.E., 2009a. The dual nature of ecosystem dynamics. Ecol. Modell. 220, 1886−1892.

Ulanowicz, R.E., 2009b. A Third Window: Natural Life beyond Newton. Templeton Foundation Press, West Conshohocken, PA, 224 pp.

Ulanowicz, R.E., Puccia, C.J., 1990. Mixed trophic impacts in ecosystems. Coenosis 5, 7−16.

Ulanowicz, R.E., Goerner, S.J., Lietaer, B., Gomez, R., 2009. Quantifying sustainability: resilience, efficiency and the return of information theory. Ecol. Compl. 6, 27−36.

Van Breemen, N., 1993. Soils as biotic constructs favouring net primary productivity. Geoderma 57, 183−211.

Van de Vijver, G., 1998. Internalism versus externalism: a matter of choice? In: Ǻugowski, W., Matsuno, K. (Eds.), Uroboros or Biology Between Mythology and Philosophy. Wydawnictwo, Poland, pp. 295−306.

Vernadsky, V.I., 1998. The Biosphere. Copernicus, New York, NY. ISBN 9780387982687. An annotated edition based on Vernadsky's original book of 1926.

Von Uexküll, J., 1926. Theoretical Biology (D.L. Mackinnon, Trans.). Kegan Paul, Trench, Trübner and Co., London.

Wackernagel, M., Schulz, N.B., Deumling, D., Linares, A.C., Jenkins, M., Kapos, V., et al., 2002. Tracking the ecological overshoot of the human economy. Proc. Natl. Acad. Sci. U.S.A. 99 (14), 9266−9271.

Walls, L.D. (Ed.), 1999. Material Faith: Thoreau on Science. Houghton Mifflin, New York, NY, 120 pp.

Watts, A., 1959. Beat Zen, Square Zen, and Zen. City Lights Books, San Francisco, CA.

Yaron, P., Walsh, M., Sazama, C., Bozek, R., Burdette, C., Farrand, A., et al., 2000. Design and construction of a floating living machine. In: Cannizzaro, P.J. (Ed.), Proceedings of the Twenty-Seventh Annual Conference on Ecosystems Restoration and Creation. Hillsborough Community College, Plant City, FL, pp. 92−101.

Index

Note: Page numbers followed by "*f*" and "*t*" refer to figures and tables, respectively.